Dennis Osborne (1927-2005)

This book is dedicated to the memory of Dennis Osborne. Without his generosity in giving me my start in model engineering, these experiments and therefore this book would probably never have happened. A great regret I shall always have, is the lost time we could have spent together.

Compiled and written by Colin Peck

Published by Ambatex Publishing 2006
Unit 7, Cranham Farm, Upminster, Essex RM14 3YB, England.

Copyright © 2006 Colin Peck

All rights reserved. Apart from any fair dealing for the purpose of private study, research, criticism or review, as permitted under the Copyright, Designs and Patents Act, 1988, no part of this publication may be reproduced, stored in a retrieval system, or transmitted in any form or by any means, electronic, mechanical, photocopying, recording or otherwise, without prior written permission.

All enquiries should be addressed to the publisher.

Although every care has been taken to ensure the accuracy of the information contained in this manual, no liability can be accepted by the author or publisher or suppliers for any mechanical malfunction, damage, loss, injury or death caused by the use of incorrect or misrepresented information, omissions or errors that may have arisen during the preparation of the manual.

ISBN 0-9546895-1-8

10 9 8 7 6 5 4 3 2

Photography by Colin Peck

Printed in Great Britain by Ambatex Publishing.

THE ARTFUL BODGER'S IRON CASTING WASTE OIL FURNACE

Contents

Introduction	1
The basic design and how it works	2
The air supply	9
Building the furnace	13
Insulation	38
The pipe work	42
Repairs	48
Using the furnace	50
Crucibles and metals	58
Casting	69
The fuel valve	72
The fan housing	79
Cast iron crucibles	83
Useful information	85

Supplement to page 52 overleaf

Supplement to page 52 on using the furnace

Curiosity finally got the better of me and I decided to see how big a crucible of cast iron the furnace could handle. I bought an A20 crucible (aprox. 60 lbs. of cast iron) as this seemed to be the largest size which would fit and still be able to be worked. I was very pleased when 2 hours after light up (this included skimming and topping up the pot three times), I poured a nice fluid pot full of metal, however it brought to light something I feel needs a better explanation than I give in the book.

The furnace burns oil on the basis of a cycle, starting up the wood (or propane) fire starts to heat both the furnace interior but more importantly the pre-heater, after 4-5 minutes the oil is turned on, now the pre-heater really comes to life. Wait for the wood to burn down (or the propane is turned off) then adjust the fuel/air flow to give you a full yellow exhaust flame with a small amount of orange in it. Remember, the idea is to fill the furnace with burning fuel with enough excess to give a good pre-heater flame.

Now the cycle kicks in, as the furnace heats up the pre-heater flame gets hotter, this in turn heats up the air/fuel mix more allowing it to vaporise easier and burn faster, this means the interior of the furnace now gets hotter, which means the pre-heater flame also gets hotter, making the air/fuel mix hotter etc. As this process is in action, the exhaust flame and the pre-heater flame get lighter in colour, smaller and thinner, also you will be able to see the interior of the furnace more clearly through the exhaust port as the fire in it shrinks due to the fuel burning so much faster. This means we can now turn the fuel/ air feed up to fill the furnace to overflowing again which keeps this cycle going and the temperature in the combustion chamber just keeps going up until you reach the temperature you wish to work at.

Don't go running this flat out with a smaller crucible, even for cast iron, it's simply too powerful and you stand a good chance of burning/melting the crucible away!

However with a larger crucible we need to generate enough heat to do the job, but we now have much less space in the combustion chamber for burning fuel to do this but a greater mass to heat up! This means that the increases of fuel/air must be done **gradually** to allow enough time for the whole of the furnace interior, the crucible and its contents to build up to the full heat possible from the last fuel/air increase, then do the next increase. Try to go too quickly and you will slow the whole process down by running too rich, also smoke and then oil will escape from the bottom of the furnace making a mess (have a drip tray handy just in case).

With a smaller crucible there is so much fire in the combustion chamber, it can tolerate some extra fuel if things are rushed a bit as it won't take long for the furnace interior and pre-heater to catch up. Since you will be learning to use this furnace using a smaller crucible (I hope!), you shouldn't have too many problems, and hopefully this page will help you when you get more ambitious.

With an A8 crucible of iron, I usually end up with my air valve around ¾ closed (remember yours will vary from mine) with the fuel flow set as described above, this gives me a furnace full of fire and generates plenty of heat to do the job. However with the A 20 I eventually had the air valve fully closed (maximum air flow to the furnace), and although the exhaust flame was higher than I would normally have it, the majority of the heat was being generated internally heating the crucible and metal.

I am still amazed that by the simple inclusion of the pre-heater hole (and its associated pipe work) waste oil can be burnt to generate this much heat within the furnace, this is the **main** reason my furnace does what it does so simply. I find it far easier to have this burning well for iron casting than I ever did with propane, (and just a little cheaper!) but maybe that's just me. Read page 52 and combine that information with this and you should be fine. Don't be scared to experiment yourself though, don't worry there's nothing to blow up!

Colin Peck (The Artful Bodger)   www.artfulbodger.2ya.com

Introduction

I've had enough, I've spent too long experimenting with ways to melt metal and not enough time working on the long list of projects that I want to do. I have achieved my goal of melting cast iron for little or no costs so I am handing over my thinking hat to any of you who wish to go further. The furnace set up described in this book does all I ask of it, if you want more, feel free to use my ideas as a starting point to greater things, alternatively build this unit and get on with your own projects. Either way enjoy yourselves, that's what a hobby is all about.

I cannot claim any expertise on the subject of foundry work; I am strictly an amateur with a very limited knowledge on the subject. This book gives a description of how I built and use my furnace, if you wish to copy what I have done then on your own head be it!

I haven't invented or designed anything new, simply put together some well used principles and concepts by trial and error until by the law of averages, I had to get something right!

Handling molten metal can be a dangerous past time if sensible precautions are ignored. While I shall try to point out as many hazards as I can, ultimately you are responsible for your own safety. The occasional sound of some cursing due to a minor burn is par for the course in home foundry work, whereas the siren of the ambulance carting you off to casualty is not something any of us want to hear!

About three years ago I decided that the ability to make my own castings would greatly improve my metal working abilities, having no knowledge at all on how to go about it I bought a book. This was "The Charcoal Foundry" by Dave Gingery, the starting place for many home foundries. If a book has the name Gingery (Dave or Vince) on the cover, and is on a subject you are interested in, I would be surprised if you feel you have wasted your money!

Deciding that the ability to cast iron was needed, I started to experiment with other fuels including coal, coke, propane and natural gas. The ease of melting aluminium spoils you, getting the furnace hot enough to melt cast iron takes some doing and very deep pockets! This presented a problem as my foundry budget consists of what I can scrounge from the household budget before my wife catches me. While I can weather the domestic storm to cover occasional overspends, regular deep dips into the domestic purse for fuel would spell the end to domestic bliss!

This book describes the end result of several years of blood, sweat and tears (and one or two burns). In the end it was all worth it (I think!).

Colin Peck

The basic design and how it works

I have built this furnace primarily to burn waste oil, old engine oil and cooking oil as this is free! You can always try any other types of waste oil that seem suitable. It will also burn most types of solid fuel and propane, as well as new oil such as diesel or paraffin. Given a choice I prefer to burn engine oil as it seems to get hotter than cooking oil, although this will still melt cast iron. Also cooking oil must be left to stand for a few days once you get it home for the contaminants to settle and then the clean oil poured off the top, whereas engine oil is ready to go. There isn't that much to choose between them though. The contaminated oil remains from the cooking oil is usable by putting some wood blocks (pre-cut to size) in a bucket with a weight to hold them down. Then pour the oil in the bucket to soak into the wood, top it up as it soaks in, and then burn it. Watch out! It burns with an enormous exhaust flame.

The ability to burn wood is very useful when a lower heat is required, such as when melting zinc alloys or smaller quantities of aluminium, no exotic hard woods here good old pine off-cuts work fine!

I'm not going to get too involved in any aspects of oil burning except where they are directly related to this furnace as there are many variations. All that must be remembered is that oil doesn't burn, only the vapour or gas produced when it is heated sufficiently burns.

We are using waste oil which is not as easy as using nice clean oil as the contaminants cause blockage problems. Heat up engine oil and the oil vaporises leaving behind carbon deposits which will block any pipe in no time flat. The debris in waste oil means any jets used would be constantly blocked, as would any filters used to prevent the jets from blocking. Cooking oil is easier to vaporise; however as a fuel turns to vapour it expands causing the oil flow in the pipe to stop until the pressure created drops. This gives an intermittent flow of vapour to the burner unless the oil is pressure fed, this then needs a jet to reduce the flow which means a small hole to get blocked!

This system doesn't have a burner, it uses an oil/air delivery system (a bit like a carburettor) which delivers pre-heated oil and air to the combustion chamber (the interior of the furnace), where it is burnt as quickly as possible. The idea is, instead of having a burner which blasts a flame into the furnace, we fill the furnace to the point of overflowing with burning fuel, and we use the "overflowing fuel" to do the pre-heating. In the middle of this inferno we have the metal we are melting.

Since we are using the interior of the furnace as a combustion chamber, the larger the crucible we put in, the less combustion space we have for burning fuel to melt it. This means we will come to the stage where we simply can't generate enough heat to do the job.

Only with cast iron does this become a problem as we still have plenty of heat available for lower temperature metals, with aluminium it's more or less a case that if the pot fits in the furnace, the metal in it will melt. The largest clay/graphite crucible I feel comfortable handling on my own is a no.8 which holds around 28lbs. of brass, or around 24lbs. of cast iron. If you add the 6lbs. weight of the crucible, handling 30lbs. of white hot stuff isn't to be taken lightly.

I do feel the furnace could handle a larger crucible, but I haven't tried it! The internal diameter of the furnace detailed in this book is 10", increase this to a little over 14"dia. and you will double the internal volume allowing for a much larger crucible and still have enough combustion space to generate the heat to melt its contents.

I haven't made this larger version, but I can't see why it wouldn't work. The only change I would make from the 10" version would be to increase the diameter of the air pipe from 2" to 2 1/2".

If you want to go larger still, I would look at having two or more inlets with separate blowers and fuel pipes; this might work better than a large single blower and massive pipe work. The blowers could be controlled with a single variable resistor, and a single fuel valve with a large capacity splitting into two or more feeds, however this is getting beyond the needs of the average home foundry man.

As a very rough guide, running hard this furnace burns somewhere around 10-12ltrs. an hour with engine oil, more with cooking oil. The oil flows from an oil tank by gravity through 15mm (1/2") standard plumbing fittings and garden hose pipe (a suitable flexible oil proof pipe would be better), to a 15mm ball lever type on/off valve, this is not sensitive enough to control the oil flow so next we need a means of doing this. In my early trials I simply used a small G clamp to squeeze the hose pipe, but there is an easy to make control valve described in this book. From the control valve we use 10mm (3/8") micro bore plumbing pipe which will enter the air pipe and run inside it.

The air is supplied by a suitable fan or blower into a 2"dia. thin walled steel pipe (stainless steel would be longer lasting), the last 1,1/2" of pipe tapers down to 1,1/2". This cone we have created causes the air flow to build up within the pipe and exit the 1,1/2" hole at a far greater velocity than it would from a straight pipe, although at a lower volume (the same effect you get when you get when you put your finger over the end of a hose pipe to make the water spray harder).

As the oil pipe is running inside the air pipe the air stream has to separate to pass along it, when the air reaches the end of the oil pipe it will flow together again. The faster the air flow, the longer it will take for the air stream to join together, leaving a cone of low pressure or vacuum at the end of the oil pipe. This causes the oil to be sucked out of the pipe and drawn into the air stream, with both pipes ending at the same length we get the benefit of the faster air flow from the cone to suck the oil out, and then as the air stream leaves the confines of the pipe, it will spread out drawing the oil with it. This causes the oil to break up into small droplets which will vaporise very easily when sprayed into the hot combustion chamber.

To burn oil in the quantities we need to obtain high temperatures means we need to supply a great amount of air to the combustion chamber.

Now if you make yourself a cup of tea or coffee it is too hot to drink straight away, so we blow across the surface of the drink to dissipate the heat allowing us to take a much needed sip.

By blowing large amounts of air into our combustion chamber we will dissipate the heat we are trying to build up in the same way! Since our oil droplets are carried in this cold air stream, on a cold day they simply won't get hot enough to vaporise.

We need to heat the air as hot as we can before it comes in contact with the oil or the combustion chamber, this way the hot air passing along the oil pipe will heat the oil inside allowing it to break up easier as it comes out of the pipe. Also if the air is hot enough, the oil will start to vaporise as soon as the two start to mix, this means instead of cold air and oil drops entering the combustion chamber, we have hot air and oil vapour ready to burn. The hotter the furnace is running, the hotter the pre-heater will be.

Any fuel takes a certain amount of time to burn; given that we are trying to burn the largest quantity of fuel within a confined space as possible, we also need to burn it as fast as possible. Simply filling the furnace with fuel burning at a slower rate will not be enough to melt iron, however with this pre-heating system we can burn a greater quantity of fuel in a shorter space of time, which means we can feed even more fuel into the furnace to keep it full, generating huge amounts of heat. Without the pre-heater, if we fed the same amount of fuel into the combustion chamber, a large amount of it wouldn't have time to burn and it would merely rush out with the exhaust gasses creating a very impressive exhaust flame while adding no heat at all to the combustion chamber, providing it got hot enough to vaporise and burn in the first place!

The air pipe enters the furnace at a tangent to the inside circumference, causing the air stream to create a vortex spiralling around the combustion chamber. We build the furnace with an outlet hole (the pre-heater hole) on the same centre height as the air pipe and at the correct angle for the flame that blasts out of this hole to hit the air pipe around 3"- 4" away from the furnace side, to concentrate the heat here we put a simple heat shield arrangement at this point. To further increase the heat transfer to the air stream, a 3/8" slot is cut through the sides of the air pipe. This allows a short piece of thin walled pipe to be flattened (to 3/8") and inserted in the slot, one end directly in the path of the pre-heater flame, this end is cut to a suitable angle and flared to encourage the some of the flame to pass through it. As the air stream splits to pass either side of this pipe, it is forced into greater contact with the heated surfaces. Lastly the edge of this flattened pipe nearest the furnace has two 1/4" slots cut/filed from it; this is to encourage hot gasses to be drawn into the air stream further heating the air.

This set up isn't very sophisticated, just three bits of pipe (oil, air and flattened) and a hole (the pre-heat hole), the smallest pipe diameter is 10mm so it doesn't block very easily, a forth pipe added (shown in the building instructions) will allow propane to be burnt.

We light a wood (or propane) fire in the furnace to start things off, after a few minutes we turn on the oil. Once the furnace has been alight for five to ten minutes the air and fuel can be turned down if full heat is not needed, leave this running flat out for too long and it's possible to reach temperatures higher than a clay/graphite crucible is capable of withstanding, I have also melted the 1700deg.C.(3092deg.F.) lining in the furnace on several occasions during the testing of this unit, beware it didn't take that long!

The furnace itself is a simple design which evolved to suit my needs, it's on wheels to make it mobile, mounted on a triangular chassis made of angle iron, three wheels allow for stability on uneven ground. Built off of this chassis are two cylinders, the inner one is a high temperature refractory lining 1,1/4"-1,1/2" thick with a 10" inner diameter which we cast in place, we then have a 1,1/4"-1,1/2" gap which will be filled with insulation, held in by a steel (or stainless steel) cylinder which is the outer skin.

We make a lid from the refractory material which has a steel outer skin to allow us to bolt it to the lifting arms, the lid assembly lifts and swivels to open via a foot pedal and handle. The bottom of the furnace is a steel plate which clamps into position with a sand/clay mix rammed on the top of it for a refractory skin, this allows the bottom of the furnace to be quickly and cheaply replaced when it becomes contaminated.

There is a tap hole and a spout built into the furnace at the level of this bottom which allows any metals (unfortunately not cast iron), which are directly melted in the furnace to be tapped out. The bottom of the inlet port is 2" above the furnace bottom creating a 2" deep well for the molten metal to collect in before tapping. This well can not be too deep as we are only relying on reflected heat to keep the metal hot enough, the bottom is the coldest section of the furnace.

The refractory lining must extend to the outer skin in two places, the tap hole and the area where the inlet port and pre-heater hole are situated; we build in the formers for these so that these extensions are cast as part of the refractory lining.

The oil burning set up can be fitted to almost any design of furnace as long as the inlet port is at least 2"dia. and the pre-heater hole can be incorporated. The only other requirement that needs consideration is an effective sealing gasket between the lid and the furnace body or any other joints in the combustion chamber. Despite having a good sized exhaust port in the lid, the expansion of the gasses that occurs during combustion creates a pressure build up inside the furnace (don't worry nothing will explode, it's not that high!) causing vaporised, but unburnt fuel to leak out as a black smoke. This is more of a problem when melting cast iron as we are feeding a lot of fuel into the combustion chamber to generate enough heat.

I call this design a top loader, there are a lot of more sophisticated designs that I have seen on the internet, some are so well put together and beautifully finished, I would be more inclined to use them as an ornament in my living room than get them dirty melting scrap in them!

There are one or two thoughts I would like to share with you concerning furnace designs and casting iron as opposed to casting aluminium.

Firstly the down-sides of a top loader, heat rises, therefore all the crucible work involved with melting iron is carried out in this rising heat, long tools are needed for this. When the metal is ready to pour, it is lifted vertically out of the furnace with lifting tongs and then the crucible must be transferred to a pouring shank or pouring tongs. However it is the easiest design to build and to incorporate insulation, also these down-sides can be looked at as "up-sides" if you flip the coin over.

What do I mean by crucible work? Cast iron produces a lot of the worst slag you will come across, the older or poorer the iron, the greater the quantity of slag, also the more contact the iron has with the furnace atmosphere the more it will oxidise and turn into slag. This can be reduced and a better quality iron produced if a lid is used on the crucible. Iron slag is like semi-fluid tar, it can hold pieces of iron suspended above the fluid metal, and is best removed before topping up the pot. It needs to be prodded and stirred with a steel poker, then scooped off. Depending on how much slag your iron is producing and the size of your iron pieces, it might be necessary to de-slag and top-up three or more times to get a brim full crucible of iron. Actually it's worth leaving a little in the pot after the final top-up to keep the air away from the metal while it heats up to a pouring temperature, but when you are ready to pour it all must be skimmed off.

Slag inclusion is one of the greatest problems with pouring cast iron, nothing worse than breaking open a mould and finding a lump of slag right in the most important part of the casting, sod's law dictates it will never be somewhere it doesn't matter!

You might not want to do this crucible work in the heat of the furnace, but while the crucible is inside it is kept nice and hot. When the lid is open the hot air rises, because that's what hot air does! However not much cold air can rush in to take its place as there are only the air inlet and pre-heater holes to let it in.

This isn't the case with a lifting body furnace, cold air will rush all around the crucible as soon as it's opened, not only will the metal cool off quickly, there is always the possibility of thermal shock cracking the pot. The furnace body itself can also suffer as cold air will be rushing up through the open bottom, also all this heat loss must be replaced.

Where is the furnace body when it's lifted? Remember it will be radiating out around 1500deg.C. (2732deg,F.) I hope your head or any other part of your anatomy is nowhere near it as you attend the pot! If your design has a lid to open as well as the lifting body you can work through the open lid, though it now sounds like a top loader with lots of other bits of metal attached to it for no purpose.

As for removing the crucible from the furnace, it's a lot easier to vertically lift a heavy weight with a pair of tongs than it is to hold 20 or30lbs. of white hot clay pot in a pair of tongs some distance away from your body, and possibly have to manoeuvre it through a restricted space. Once you've done what you need to do, it then has to go back in the furnace ready for the next time.

The lifting tongs I use can only grip the pot in the right place so there is no fiddling around, it only takes a couple of seconds from opening the lid to placing the pot down for it's final skim and pour. We are not melting enough metal in home foundry work for the mass to retain the heat for that long once the furnace is turned off, we must work quickly and efficiently to get the best results, but most of all safely!

With a simple design there's very little to go wrong, I don't want to stop you building the furnace you think is right for you but remember, if you build a furnace capable of melting cast iron, sooner or later you will melt cast iron, and that's not the time to find out you've got it wrong! Give it plenty of thought beforehand.

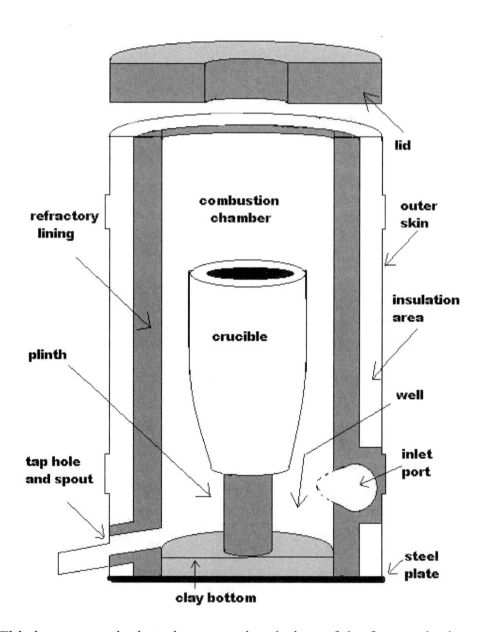

This is not to scale, but gives a sectional view of the furnace body. On page 27 there is a small drawing showing a top view giving the general layout for the inlet port, pre-heater and the tap hole.

The air supply

We need a reasonable volume of air but not at a great pressure, oil burning requires more air than burning solid fuel or gas.

The simplest and probably most available device to use is a vacuum cleaner, a lot of cylinder types have a hose fitting on the air outlet so from our point of view they are ready to go. On most cleaners the air passes over the motor windings to keep them cool, so the air arriving to our system already has some pre-heating.

I don't suggest you use the household cleaner if you want to keep things peaceful on the domestic front, whatever we use will get messed up and oily. I have picked up a number of old cleaners from outside neighbours houses on rubbish day, they all know I have some strange habits! I've been lucky, every one has worked, they simply had blocked filters or broken parts that I didn't require anyway.

Shown are some pictures of an upright cleaner that doesn't have a hose fitting on the outlet. When all the unnecessary parts are removed it leaves a compact motor and fan unit.

Remove any filters and cut a hole in the outlet grill slightly smaller than the hose, then pushing the hose in this hole gives us a blower. The hose needs to be glued in and the rest of the grill filled in so all of the air exits through the hose, I had some old car body filler which did the job nicely. Picking up a dumped cleaner and about an hours work to create a powerful blower is good enough for me!

Since the motor cooling is provided by the air flow, I didn't want to restrict the air flow to adjust the volume supplied to the furnace. If you are happy to mess around with electrics, a good system is to use some form of variable resister similar to the variable speed control on some electric drills. However the one I tried burnt out in a short time, cleaners have powerful motors, and I only use this example to explain the type of unit to look for, make sure it's rated high enough for the power of your motor. I'm actually using a household light dimmer switch for my home made fan unit shown in this book, it's probably not rated high enough for the 1/10 hp. motor I'm using but it works, so far so good.

I made this fan unit because I had the motor (a 2800 rpm. ex-spin drier motor), the light dimmer switch, some bearings and a hover mower fan; I was curious to see how well it would work. I give some of the details of building this later in the book, but since I built it using what I had (most parts are cast), and used the lathe to make the fan bearing unit, some people might not be equipped to build it. If you look at the information I give and have, or come across the required parts, then you should be able to put something together, if not, the casting of the fan housing should give you some ideas for a quick moulding method for other projects. If I didn't have the parts I would be using a cleaner/blower.

I show in the pictures a simple air valve made from a 1,1/2"compression plastic waste pipe T, this is the size used for baths and kitchen sinks. The way the pipes are fitted, the air flow will mostly blow out of the back; the shutter variably blocks this exit adjusting the amount of air diverted to the furnace, without restricting the amount of air passing through the motor, remember, keep it simple!

One problem with a cleaner/blower is they are plastic with plastic hoses, where the hose connects to the metal air pipe the air flow will keep every thing cool. But when the air is switched off, heat can travel up the metal pipe to melt the cleaner hose. It's necessary to use a longer piece of metal pipe to keep the plastic parts away from the heat, also pulling the air pipe out of the furnace when the blower is off, later in the book I will explain why this is a good idea anyway. It is possible of course to connect a metal pipe directly to the cleaner, if the pipe is welded to a suitable sized plate, this could be screwed over the outlet grill with some sealer around it. The flexible pipe does make pulling the air pipe out of the furnace easier though.

Any hot embers or sparks will make holes in the plastic, the motor should be far enough away not to come to any harm especially if it is in a box, but the hose can suffer, although the longer metal pipe keeps it away from too much harm. A roll of sticky tape will keep things going, and when it get too bad, look at your neighbours rubbish for a replacement hose.

Another problem is noise, for some people this won't be an issue, but these things are noisy. Bear in mind the furnace itself isn't exactly whisper quiet to say the least!

Let me introduce you to "the hoover in a box". This is the motor unit shown, placed in a rough plywood box stuffed with fibreglass loft insulation. There is a hole with the hose coming out, and I made a rough inlet tube from some thin aluminium sheet just bent to size with my hands which gives a clear passage of air from outside the box to the hoover inlet. A couple of 1/2" square pieces of wood space off an off-cut of ply to cover this inlet, allowing the air to enter only from the top and bottom, helping to reduce the noise further.

You can't get a lot simpler than this, it works perfectly, and when it eventually packs up it's not difficult or expensive to replace!

The "hoover" blower with the hose fixed into the outlet grill with car body filler

The redundant parts to the left, with the "hoover" blower to the right

The "bath T" air valve, note the shutter screwed to the back with a self tapping screw

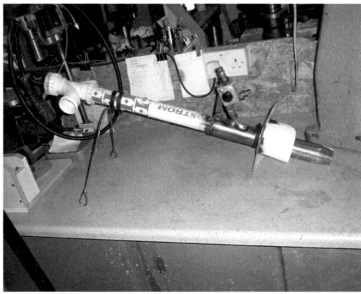

The complete air pipe assembly including the fuel valve and the air valve. The red and white pipe is to extend the stainless pipe and is a smaller dia. which helps it to fit in the plastic T.

Condensing tumble driers have not one but two centrifugal fans running off a high rpm. motor, cut off any unnecessary bulk and you have a powerful, quiet running blower. It must be mentioned that reducing the outlet of a fan unit to the size of the air pipe must be done with a gentle taper, centrifugal fans produce a high volume of low pressure air, too much resistance or turbulence will kill the air flow.

What else can we find lying around? Perhaps a leaf blower, I haven't tried one I'm just suggesting some possibilities. A car heater fan? You could use the variable speed control from the car and power it with a battery charger, or use a light dimmer switch to vary the supply to the battery charger as a volume control. I mentioned a hover mower fan earlier, I used the motor for something else, however if you have an old mower gathering dust in the shed you will have something to start with, using the cast fan housing I will describe later in this book, I'm sure something can be worked out, don't forget to remove the cutting blade!

Look around you, what's lying around waiting to be adapted? One thing to bear in mind, we do require a certain volume of air for the furnace to get hot enough to melt cast iron, I can't give a lot more advice than to try it and see. If it doesn't provide enough air for iron, you could use it for all other melting and use a cleaner purely for your iron work. Or can you "gear up" by belt driving between the motor and fan as I have on my blower unit. It can start to get very involved, when a simple cleaner will do the job well (unless you have found an incredibly useless cleaner), and reasonably quietly with the "Hoover in a box" set up.

Building the furnace

It is very difficult to describe with words alone how to build something, so I have included as many pictures and drawings as I can to help. Read the text as well though as I have modified some things after some of the pictures were taken, I will try to point these out as we go along.

Hopefully you will find a lot of suitable bits and pieces knocking around to build your furnace, and as not everything will go together in the same way as my bits and pieces did, a little bit of artistic interpretation on your part will be needed, so I try to explain any areas I feel are important, in enough detail to allow for this. Just remember that it's a very simple concept, not very accurately put together.

Read through all the build instructions before you start to make anything to get an idea of what you will need and if what you have will be suitable for the job. Jot down some notes of anything you will be changing or will need to modify, I usually remember half way through doing it wrong!

I make no excuses for the appearance of my furnace; it was built with what I had or could obtain cheaply, except for the refractory lining. I wanted this to be the last furnace I would build for some time, so I paid out for a commercial 1700deg.C.(3092deg.F.) castable refractory cement. This stuff goes off chemically like concrete and is almost as hard. 1600deg.C.(2912deg.F.)is also o.k. to use, and is sometimes easier to find.

It comes in 25kg.(55lbs.) bags and isn't particularly cheap. I bought 3 bags and had almost a bag left over, it's worth having too much as you will need some to make the plinths for the crucible to stand on. The lining thickness of my furnace is 1,1/4". I would have liked 1,1/2" but the beer barrel I used for the outer skin left me a little short on space if I wanted to keep to the 10" bore I had used on my last furnace, three bags would have still been enough for this extra thickness.

I had to have this delivered, which cost almost as much as another bag of refractory! There is a list of companies I have dealt with in the back of the book, but search in your own area, you might find it cheaper or at least save on the delivery charge.

If you look on the internet you will find various blends for homemade linings, these are mainly based on fire clay, which is something worthwhile to track down anyway as we can use it in this furnace for a couple of jobs. There are ways around using fire clay, but it does work well. With these homemade linings, some have insulation included in the mix; I don't recommend this as most of these insulating materials cannot withstand the kind of heat we will be generating.

I am going to assume that you will use a commercial refractory for this construction, however if you want to try a homemade mix it will certainly be a lot cheaper, and to re-line the furnace is not too major a job.

This construction involves welding, if you don't weld you should be able to make the oil burning set up and incorporate it into a simple refractory lined, drum type furnace, there are plenty of examples of these on the internet, look at what I've built and see if you want to include anything I have shown into your design. I list some home foundry web sites in the back of the book; look through these and their links for ideas.

The outer skin I've used is a stainless steel beer barrel, another inch in diameter would have been nice but apart from this it's ideal. It measures just over 15"dia. by 22" high. If you want to make a larger furnace, then adapt the measurements given to suit your size. I don't recommend increasing the height, remember you will be lifting the crucible out of the top and too high will be dangerous.

It's best to find a strong barrel or drum of a suitable size with a rolled or otherwise finished top rim, or one with a fixed top that you will need to cut out, this way you can leave 1/4"-3/8" of the top in place for strength. This skin needs to be quite strong and to be able to hold it's shape, the bottom will be welded to the chassis, but the top needs to support itself.

If you do make the skin yourself, use the thickest steel sheet you can work with.

Bend to shape and weld a circle of steel bar (5/16"-3/8" square or round) inside near the top to strengthen it up. You can use the same method described to make the removable inner former for all three of the cylinders you will need, simply weld the outer skin and the outer former to form the cylinder.

Stainless steel is a lot harder to work with than mild steel, it will weld (electrically) with ordinary steel rods or wire, but is hard work to cut or drill. The top of this skin must be level as we use this to level off the lining and the insulation capping.

The beer barrel has raised sides above the level of the top with handle holes; there is a nice indent all the way round the barrel which you can follow for a straight cut. Using an angle grinder it's possible to cut off this side extension (this will be used for the lid) and the top in one go. Give any cut edges a good clean up after cutting as there will be some very sharp burrs left!

All other cutting including cutting the bottom out was done using an arc welder with the amperage cranked up to burn the metal away, not very neat, but nice and quick! You can buy cutting rods for this purpose but I just used some old welding rods, otherwise use whatever means you have.

Firstly we will need the outer skin, the outer former and the removable inner former; these formers are made from thin steel sheet (about the thickness of car body work or thinner).

With the top cut off of the barrel these will be just under 18,1/2" high. When they are all in place on the chassis, the formers must not stick up above the outer skin as we will be using a straight edge across the outer skin to level everything off.

The inner former is 10" dia. To make this, cut two 10" dia. discs from thin scrap wood, hardboard is ideal for this; the sheet metal can now be rolled over these discs to form a cylinder. Where the metal overlaps the outer edge will want to stick out, this will need bending in a little to form a nice joint. Use a thin piece of rope tied in a circle a little larger than the cylinder, pass this over the cylinder to the middle. Insert a piece of wood through this rope circle and wind it round, this will tighten the rope around the metal sheet until the metal is firmly gripping the hardboard discs. Parcel tape or other suitable tape can now be used to hold the cylinder together and the discs in place, this will make it easy to remove this former when the refractory has set.

The outer former can be made in the same way, but the overlapping joint will need to be welded and the discs removed as this will be staying in the furnace to support the lining. On my furnace the lining is 1,1/4 thick, so the dia. of the outer former is 12,1/2"

Do you remember from your school days how to work out the circumference of a circle? Pi. (3.1416 to 4 dec. places) x dia. So the inner former metal sheet will be 31.416" wide, call it 32,1/2" with the overlap. The outer former will be 39.27" or 40,1/4" with the overlap.

Three pieces of 1,1/2" angle iron make the chassis, on mine these sides are 17" long. I didn't run the angle to sharp corners but welded short pieces across to form blunt corners.

Put the angle iron on a flat surface with the horizontal surfaces facing inwards, sit the outer former centrally on the horizontal surfaces snugly into the corners of the angle, and even up the three legs. Cut the short corner pieces to length, place them on top of the horizontal surfaces and weld securely in place, when the lining is in place it will sit nicely on the horizontal surfaces.

The outer skin can now be placed on the chassis and centralised around the outer former, mark where the uprights of the angle iron meet the skin so slots can be cut in skin to allow it to sit on the horizontal surfaces the same as the former. The inner former should sit in the middle just inside of the angle iron, mine was a nice snug fit, it doesn't matter that the inner former is slightly lower because it doesn't sit on the angle iron, this is not precision engineering!

I welded the outer former to the angle iron, but I think it is better to bolt it to the upright sections where they meet, drill and use 3, 6mm.(1/4") nuts and bolts with the bolt heads to the inside, this will allow for slight adjustment as the build progresses, however don't bolt it together yet, mark one bolt hole to the adjoining angle iron so you know where it lines up when you replace the former.

We now need to cut the holes and make the formers for the refractory lining to extend to the outer skin for the tap hole, the inlet port and pre-heater hole. Two fixed wheels are fitted to the ends of one of the angle iron sides; these will make the furnace mobile, and also act as a pivot to tilt the furnace and pour metal that is directly melted. In the middle of this side we locate the tap hole. This will be cast with a slight fall i.e. the inside will be higher than outside to assist the metal to run out.

I made my tap hole 3/4"dia. and we need around 3/4" of refractory lining all around this hole, so the former needs to be about 2,1/4" dia. The upright of the angle iron will determine the lowest height this former can be fitted in the furnace, if you used bigger than 1,1/2" angle you might want to grind some away in this area to allow for the former. Measure the height and cut the 2,1/4" hole in the outer skin so that the former will slope up about 1/8" to where it goes into the outer former and either sit on or just miss the angle iron, then cut the hole in the outer former to suit, once again this isn't precision stuff! I cut all these holes with the arc welder.

Roll a thin piece of sheet steel to make a tube to fit in these holes. Mark around this tube with a felt tip where it meets the outer former and the skin, add an 1/8" either side then cut this out with tin snips, when we are ready this can then be tack welded into position.

The inlet port and the pre-heater holes are set in one long block of refractory rather than messing around with two formers so close together. The hole in the refractory lining for the inlet port wants to be directly opposite the tap hole. Put the inner former in place and put a piece of 2x2" wood on the top of the furnace with its outer side just above the inner former, this will show the line of the holes that must be cut in the outer former and the outer skin +3/4" for the former we are making.

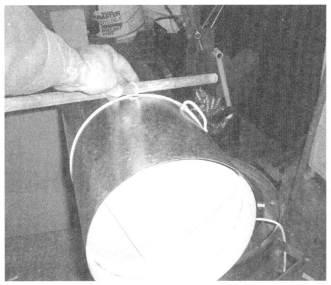

Winding the string to tighten the sheet metal onto the hardboard discs

The inner former securely taped ready for use

The outer former in position on the angle iron chassis

The slots in the outer skin allowing it to sit on the chassis

You must work out the height for this former, don't make the same mistake as me! I cut the tap hole to the size for the former and made a "guestamate" as to the height the finished tap hole once the refractory was in place. The bottom of this hole is the height for the bottom of the furnace when it's rammed up, therefore the bottom of the inlet port (not the bottom of the former we are going to make) needs to be 2" above the furnace bottom to form the well. I did my "guestamate" from the outer skin, not where the tap hole would finish on the inner former. This made no allowance for the slope of the tap hole when it's rammed up so my well was nearer 1,1/2" deep. I had to use very little depth of refractory in the bottom of the tap hole former to compensate and increase the well depth.

I made the tube for the inlet port and the pipe used for the pre-heater hole, and held them in place to get the final hole size, sort of cut a bit, try it, cut a bit more, try it etc. until I had a good position for both pipes with 3/4" clearance all around for the refractory, I think I mentioned before about it not being precision engineering! Remember, all we are doing here is making an area of the lining extend to the outside so that we can have two passages that go from the inside to the outside, with enough lining material around these passages to protect the insulation from the heat that will be passing through them.

On my furnace I didn't fit a tube for the inlet port; I used a pipe as a former and removed it when the refractory had set. I did this because on my previous furnace the pipe melted away during use leaving blobs of metal that stopped the air pipe from going in properly, I had to burn these away with a cutting torch. However a bit of clumsiness on my part with this one broke away a section of the refractory. So I fitted a guide pipe on afterwards and had to repair the refractory. It's easier to fit the inlet tube during building and if any blobs get in the way, remove them (carefully!) with a cold chisel while they are red hot after the furnace has been running.

This tube needs to be a larger dia. than the air pipe so that the pipe fits into the tube with a reasonable clearance to allow for any expansion when everything's hot, and allow the air pipe to be removed with ease at any time. A piece of air pipe can be used for this tube by cutting along it's shortest side (once we have cut the end profile) and opening the pipe up to a larger dia., then tack weld a strip along the gap carefully so the air pipe still fits inside.

It's probably a good time to make the taper on the end of the air pipe, where we bend the sections in, it tends to leave small humps which must be allowed for when checking this clearance, check the details for making the taper shown later in this section, sub headed The pipe work (P.42).

Although this is a later picture, it shows the three cylinders in place

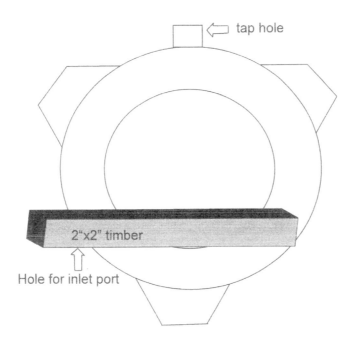

This shows the line for the inlet port

The inner end of this tube needs to be cut to a profile to fit the inner former so it all "flows" nicely; this is done using a method called Parallel line development, which will give you a nicely shaped pipe in less time than it will take me to write about it.

The drawing shown is near enough to scale to use with a 2"dia. air pipe, and the 10"bore lining described in this book. Trace and cut out the pattern, wrap it around the tube and mark the outline, then cut the tube to this profile, I did this with an angle grinder. This does not give you the length of the tube, only the end profile. It will probably be easier to leave the tube too long to make it easier to hold in position when the lining is rammed up, and then cut too length once the lining has cured. If it protrudes from the outer skin about 3/4" when it's finished it will be about right, so leave a few extra inches on for ramming up. Remember this is for a 2"dia. tube which will need to be cut in the side, to expand the dia. to allow the air pipe to fit inside.

If you are using any variation of sizes, either in the lining dia. or the dia. of the inlet tube (you might have a piece of pipe of a suitable dia. to make the inlet tube without having to cut the side, or are using a larger dia. air pipe), then it is easy enough to make a new pattern. I drew this pattern on lined paper so all of the horizontal lines I drew were easy to draw parallel using the printed lines on the paper as a guide for the ruler.

Using a compass, draw a part circle the dia. of the furnace bore, in my case 10". From the top of this part circle, draw a horizontal line that represents one side of the inlet tube (line 1-1), measure down from this half of the tube dia. in my case 1" and draw another line (3-3), then measure down again to the full dia. of the tube, in my case 2" and draw this line (5-5). Now draw the vertical line A-A at a right angle to the other lines, this needs to extend up high enough to be part of the pattern we are creating. Set the compass to the radius (half the dia.) of the tube you are using, place the point of the compass where line 3-3 intersects A-A and draw a half circle which joins up with lines 1-1 and 5-5.

Leave the compass at this setting and place the point where A-A and 5-5 meet and draw a small arc outside of the half circle, move the point to where 3-3 intersects the half circle and draw another arc to intersect the first arc. Place the edge of a ruler at the intersection of 3-3/A-A to pass through the intersection of the two arcs and draw a line, this gives us a half way point on the half circle between where 3-3 and 5-5 intersect it. Draw a horizontal line from this point to meet the 10"dia. part circle giving us line 4-4, using the same method draw in line 2-2. Lines 1-1 through to 5-5 should all be parallel.

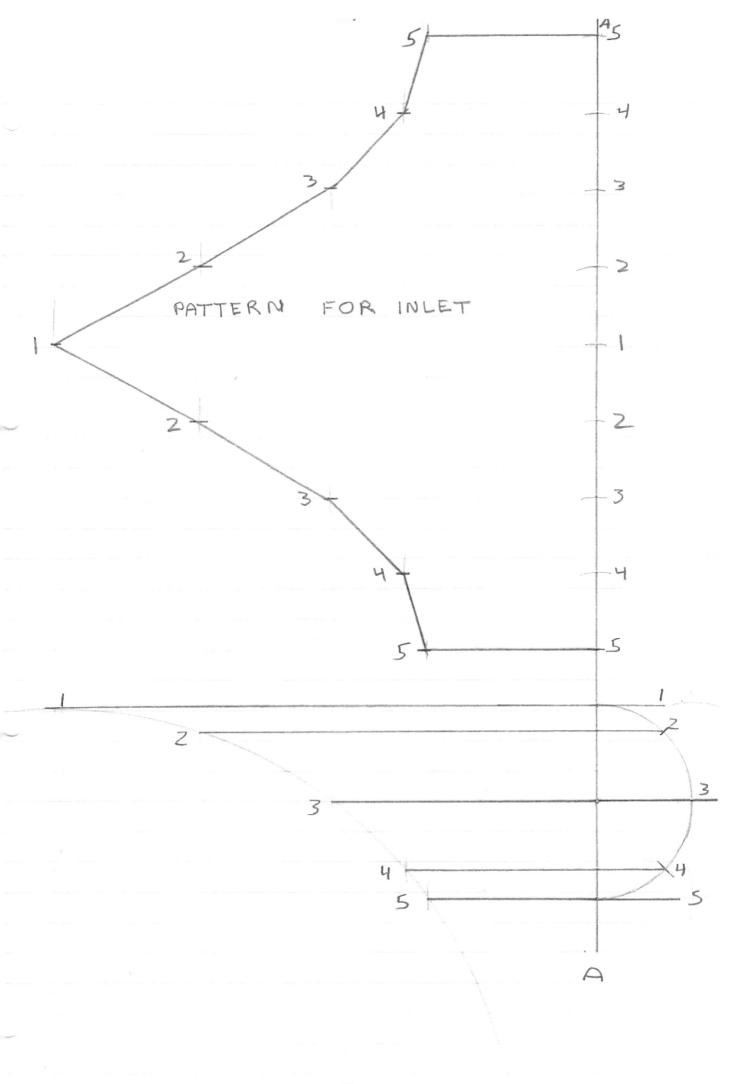

Place the compass point where line 3-3 intersects the half circle, and the pencil point where 4-4 intersects the half circle, we will use this setting to mark equal spaces on line A-A.

Leave about a 1/2" gap above line 1-1 and mark line A-A at this point, place the compass point on this mark and draw another mark on line A-A where the pencil touches it, now put the point on this mark and make the next mark etc. until you have 9 marks on line A-A.

Place the point of the compass where A-A intersects line 5-5 and the pencil where 5-5 intersects the part circle. Transfer this distance to the pattern by placing the point of the compass on line A-A at the first mark and drawing a small arc, using the same setting, place the point on the ninth mark and draw another small arc. With a ruler, draw horizontal lines from A-A to these arcs (the printed lines on the paper will keep everything parallel), these lines and A-A are the outside lines of our pattern.

Put the compass point where A-A meets 4-4 and set the pencil to where 4-4 meets the part circle and, using the second and eighth marks on A-A draw arcs in the same way, you can draw lines from A-A to these arcs, or just put the ruler in place and draw where the line meets the arc to give a cross. Do the same for 3-3, 2-2 and 1-1 which will be the centre of the pattern. If you now join these crosses up you will have the pattern to cut out and wrap around the tube. Use a little artistic flair when joining up the crosses, remember these shouldn't be a series of straight lines joined together but a gradual curve.

Instead of measuring the distances with a compass, a set square can be used to transfer the points directly from our part circle to the pattern above it. However, unless you are very careful or have a drawing board, inaccuracies can creep in.

Reading through these instructions it seems very complicated, it isn't! Look at the drawing I've done as you read the instructions and it should all become clear.

The pre-heater hole is best formed with plastic pipe, as is the tap hole, at least if they are too tight to pull out when the refractory has cured, they will burn out as soon as the furnace is lit. I used the same 3/4" plastic pipe for both holes, but wrapped and taped about three turns of cardboard from a cereal box around the pre-heater pipe to enlarge the dia. (I didn't have the right size plastic pipe) to around 7/8". The profiles at the inner ends of the plastic pipes are not as important as the inlet tube, you should be able to cut and file something pretty close. Any "bits" of refractory that are where you don't want them once the lining has cured, can be removed by carefully using a tungsten carbide masonry drill as a rotary file.

The holes burnt out for the tap hole former

The tap hole former welded in place

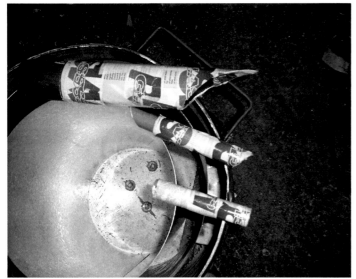

The pipe work for the three ports, the inlet port pipe is shown wrapped in cardboard as I wanted to remove it It is best to leave it in

The hole for the inlet port and the pre-heater with the cardboard pattern for the former

On my furnace the spacing between the inlet port and the pre-heater holes is around 1,1/4" on the outside and 2,3/4" on the inside. These are just how they ended up once every thing was together and is not critical. The centre of the pre-heating slot in the air pipe is around three inches away from the outer skin, although you will not be cutting this slot until the refractory has cured and you can mark the correct place to suit where your holes ended up.

When you can hold the inlet port tube and the pre-heater pipe in place against the inner former and have about 3/4" clearance between them and the holes you have cut in the outer former and the outer skin, you're about there.

To make the former you can cut four separate pieces (top, bottom and the sides) from thin steel sheet and tack weld them it place. Or I made a cardboard template from pieces of cereal box taped together.

I placed the cardboard into the holes so a reasonable amount protruded both inside and outside, I had to tape 2 pieces together to get the length. I then marked where the corners were (the folds in the cardboard don't show up very well once the cardboard is straightened up), and drew around the card with a felt tip where it met the inside of outer former and the outside of the outer skin then added 1/8". Remove this card, open it up to make a strip and cut out the pattern. Place this on a thin steel sheet and mark it out, including where the corners need to be bent and cut out with tin snips. Bend at the corners and check that it fits.

Bolt the outer former in position and put the outer skin in place, have a final check that both the tap hole holes and the inlet port/pre-heater holes still look good. Check the gap between the outer former and the skin is even all around and nothing sticks up higher than outer skin, then weld the skin securely to the angle iron.

About 1,1/2" from the top of the furnace, drill three evenly spaced 6mm.(1/4") holes around the circumference through both the skin and the outer former, these are so we can bolt the two together for support. From the inside of the outer skin pass 6mm. bolts through the holes and fit a washer and two nuts to the bolts in the gap between the outer former and the skin. The washer and one nut are tightened on the outer former; the other nut is run out until it touches the inside of the outer skin. Another washer and nut are now fitted to the bolt outside the outer skin and tightened, clamping the skin between the two nuts. Adjust the position of these nuts clamping the skin so there is no distortion to the outer former then cut off any excess thread.

Now fit in and tack weld the tap hole and pre-heater/inlet formers, once you have a couple of tacks to hold them in place, gently rivet over the

Bending the inlet/pre-heater former using two pieces of angle iron in the vice

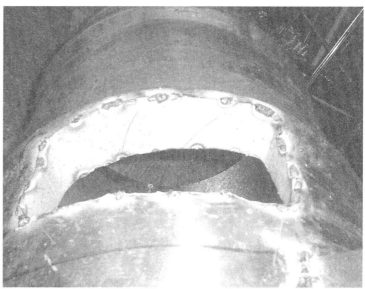

The former tack welded in place

One of the 6mm. bolt between the outer former and the outer skin

The thin plates welded in place to contain the insulation

1/8"excess inside and out, and put a few more tacks on. Don't worry if these aren't very neat, mine aren't!

The gap between the skin and the outer former will be filled with insulation, so we must weld thin plates on the bottom between the angle iron and the skin to stop the insulation falling out. Turn the furnace over and cut the small plates to size and tack into position.

I have changed the insulation from the type I originally used which was just poured in, now it has to be rammed in, so it would be best to leave the plate that covers the area under the inlet and pre-heater former off until the insulation is rammed in, then the furnace can be turned over and this section rammed in separately before either using self tapers or tack welding the plate on. This was already welded in place on mine and I would prefer more insulation in this area.

While it's upside down it's a good time to fit the wheels. The ones I used were given to me some time ago and work well, but what have you got lying around? The two wheels either side of the tap hole don't swivel around, this makes them a very good pivot point to tip the furnace to get the metal out of the spout. I originally welded them on too far back and the furnace was a bit unstable when the lid was open, if the axel is directly below the outer edge of the angle iron or further out, this will cure this problem. The width is determined by the width of the chassis. The positions of my wheels are shown in various pictures and I have drawn a sketch showing another possibility. The third wheel swivels like a supermarket trolley wheel (hint, hint) to allow steering when moving it about.

The furnace doesn't want to be too high off the ground; mine measures 5" from the ground to the bottom of the angle iron. Too high will make it more difficult to lift a crucible. This is too low however to put a ladle under the spout, so I put the furnace on bricks when I want to direct melt metal, one brick under each wheel. With the wheel located in the frog it gives enough height and is a very stable set up. Remember that there is a steel plate that makes up the bottom which will be clamped underneath; look at what I've done to make sure that your wheel arrangement won't interfere with this!

It would be a good idea to fit the steel plate now, or at least the clamp brackets and a temporary plywood plate (I used one to construct the furnace as I hadn't come across a suitable piece of metal plate at this time).

Four 1" wide pieces of the 1,1/2" angle iron are welded in place for the brackets. Two are welded to the chassis either side of the tap hole, as wide apart as the wheels will allow. The other two are welded to the outer skin on

The non swiveling wheels are fitted too far back in this picture, I re-welded them with the axels below the outer edge of the angle iron, look closely at the picture on the right for the new position

Look carefully at the two G clamps, these are where the angle iron clamp brackets are welded on to the front edge of the chassis, also note, the wheels are further forward to improve the stability when the lid is open. The lid sits over the wheel on the right when its open

The two angle iron clamp brackets welded to the skin either side of the swiveling wheel, in this picture the temporary plywood bottom is shown

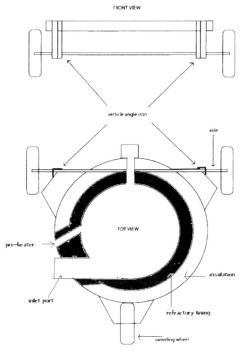

Another way to fit the fixed wheels using angle iron

the opposite side of the furnace, so that a rectangular plate can be clamped with small G clamps by its corners to the brackets.

I put the plate on a couple of bricks then placed the furnace onto the plate in the correct position with its wheels clear of the ground, next position the angle iron brackets at the corners of the plate and put a thin piece of cardboard packing between them and the plate, providing all looks good with your wheels etc. weld the brackets in place.

Turn the furnace the right way up and make a handle. I made mine from some flat bar from the scrap bin, make sure it is large enough to get a gloved hand around it. This is welded to the outer skin above the swivelling wheel about an inch or two down from the top.

The lid is opened with a seesaw type foot pedal mounted to the steel bottom plate. By treading on the front of the seesaw we push up a steel bar which slides in a tube welded to the back of the furnace; this steel bar is connected to the lid by the lifting arms. The lid lifts up and swivels round to allow access into the furnace. There is a bolt fitted near the top of the steel bar which slides vertically in a slot cut into the top of the tube. When the lid is lifted to its full height, the bolt is clear of the top of the tube. As we swivel the lid round the bolt will go out of alignment with its slot and will sit on the top of the tube preventing the lid from dropping down again until it is returned to its central position and the bolt and the slot line up again.

The bar will need to be around 3/4" -1"dia. (thick walled tubing could be used instead). The tube that this bar slides up and down in needs to have a bore which is a very loose fit with the bar. What we would consider to be a perfect sliding fit would soon jam up holding the lid partially open, a loose fit is needed. This tube is welded in a position that will allow the lid when it is open to sit above one of the wheels for stability. A look at the pictures will show you the idea.

If you haven't got a suitable skin for the lid (an off cut from the outer skin?) it's time to make one. The top part I cut off the beer barrel is about 3,1/4" high, however between 2"and 2,1/2" would be sufficient.

If you are making one, you'll need a strip of steel sheet 2"-2,1/2" high, that is formed to the same dia. as the furnace skin and welded in a circle, the thicker this steel is the better.

A lot of lids shown on the internet and those I used on my earlier designs use wire woven through the skin and embedded in the refractory to hold the refractory in place. The lids I made this way always cracked badly; I put this down to uneven expansion and contracting of the different materials

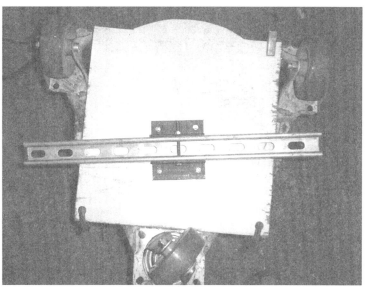

This is an overall look at the furnace, the tube welded to the skin for the lid lifting is on the right with the steel bar and lifting arms coming out of it, you can just see the foot pedal on the left

The "see-saw" foot pedal fitted to the temp. plywood bottom with two pieces of angle iron. The shaped edge of the wood is not repeated on the steel plate bottom which is rectangular

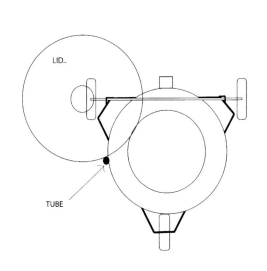

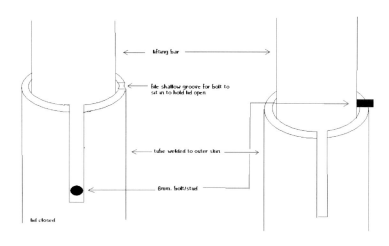

Detail of the tube showing the slot and the bolt, lid closed on the left, and open on the right with the bolt holding the lid up

Position the tube so the lid sits over the wheel when open for stability

With the off cut from the beer barrel, the lower part of the sides return inwards where it was cut off in the indent, therefore preventing the refractory from falling out, also the handle holes help a great deal.

If the bottom edge of the lid skin has two 3/8" deep slits, 3/8" apart cut at regular intervals (about every 3"), and the tags created are bent inwards about 45deg. these would prevent the refractory from falling out. We also have the two bolt heads which attach the arms to the skin sticking out on the inside of the skin which will also help.

Drill two 8mm.(5/16") holes opposite each other and half way up the lid skin. Fit an 8mm. bolt in each hole from the inside, fit a washer and nut on the outside and tighten them up. These will need to have around 3/4"-1" of thread beyond the nut to allow for a metal strip from the lifting arms with a washer either side of it and another nut to be fitted.

Either clamp or hold the lid skin to the metal bar you will be using as the lifting bar and hold this against the furnace to establish where the tube will be welded on. When the lid swings open just enough to allow clear access to the whole of the inner former, this will be its fully open position, much more and the furnace will lose stability when the lid is open in use.

The lid needs to be as central as possible over the fixed wheel which is anti-clockwise to the tap hole in the open position, mark the outer skin in this position.

Cut the tube to length, around 2" shorter than the height of the skin, when it is in place this will be an 1" short top and bottom.

Next cut a slot, 1,1/2" long at the top of the tube for the bolt to slide up and down in, I used a 6mm. (1/4") bolt, but cut the slot to whatever size bolt you will be using. I am fortunate enough to have a small mill which I used to cut the slot but an angle grinder should do the job fine, it doesn't matter if you cut the slot a little longer to compensate for curve of the cutting disc. At the top of this slot widen the entrance a little so the bolt finds the slot easily when closing the lid. Marking the position of the bolt in the bar will be done later.

Having established the position for the tube, weld it into place with the slot at the furthest point away from the skin. If you are using a mig welder, filling the gap between the curve of the tube and the curve of the skin won't be a problem, however with an arc welder you will need to bridge this gap by building up with several passes, don't overdo the heat and distort the tube or blow through and block the inside of the tube!

An alternative is to weld a piece of square bar or solid tubing (around 1/2"- 3/4" square) to the round tube first, then weld this square section to the skin, it doesn't matter that the round tube will be standing off from the skin.

his shows the tube welded to the skin with the slot in
e top, and the former for the inlet port/pre-heater

The bottom of the tube with the steel bar sitting on the
see-saw pedal with the temp bottom, before the front
two brackets were welded on

general view of the furnace, you can just see the
andle welded to the outer skin on the left

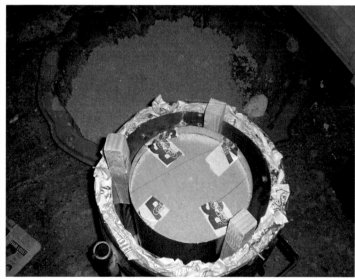

Ready for ramming up, wooden blocks keep the inner
former central with nails in them to stop them falling in.
The first bag of refractory is mixed in the plastic shell

On the beer barrel there are two rings standing proud of the rest of the skin, I just welded to these, around an 1" of weld each side of the tube near the top and bottom is all that's required. The rest of the work on the lifting arms etc. will be done once the lining and lid are rammed up.

Clamp either the steel plate or the temporary plywood plate to the furnace. Stuff old news papers or whatever in the gap between the outer lining and the skin. Put the inner former in position, then using three pieces of scrap wood make some spacers to go between the inner former and the outer former to stop the inner former going out of line during ramming up, these will be removed as the lining goes in.

The lid wants to be cast on a flat smooth surface, and we need to mark the middle of the ring for the exhaust hole. Draw a cross on the board or whatever you use to cast the lid on. You can now measure out from the centre of this cross to evenly place the lid skin in position. I tapped a few nails lightly into the board outside the skin to hold it in place.

Being "professional", I used a food tin as the former for the exhaust hole. These are about 3" dia. which works fine. Measure out from the centre of the cross to centralise the tin, I then tapped some nails to sit inside the tin into the board to hold it in position. When the refractory had set it was no problem to collapse the tin to take it out, Hi Tec. stuff!

It's very important to find out how much water is needed per bag for the refractory. The 1600deg.C. I used in my previous furnace needed more water then the 1700deg.C. I used for this one, ask the suppliers for the information but go for the maximum amount of water to make life easier. It's mixed up a lot drier than sand and cement and is a little strange to work with the first time.

It is a good idea to have a helper around when it comes to fitting the formers for the holes and the inlet tube, I managed on my own but it was a struggle, an extra pair of hands would have made life a lot easier! If I was making these for a living I would work out a jig to hold things in place, but for a one off it's not worth the effort, just rope in a friend.

Once you start the ramming don't stop until you're finished, if you haven't done it before, it will take longer than you think. I mixed up one bag and rammed it before mixing the next, using an old plastic child's sand pit to mix in, and my hands with rubber gloves on, to do the mixing. Some kind of tray is useful to put under the areas of the tap hole and the inlet port as refractory will fall out as you are working, it saves it going on the floor and can be put back in.

Tools for the job, a good ramming stick, something about 6" longer than the height of the furnace, and about an 1" in dia. The lining takes firm

The lid ready for ramming, with nails holding the can in the middle and the outer skin in place

All rammed up and setting nicely, the tape on the edge is to cover handle holes

The refractory lining in place

A view of the inlet pipe and the pre-heater

ramming, a steel bar has more impact for ramming but gets heavy after a while. A wooden dowel works but needs more effort to compact the lining. Another ramming stick, I used a wooden stick about 1/2" square to ram around the pipes. A straight edge to level off the top, a spirit level works fine, and a spray bottle full of water, the kind a lot of cleaning products come in with a trigger to atomise the contents. Lastly a few food tins to make some plinths.

The idea is to firmly pack the refractory between the inner former and the outer former with absolutely no voids. About 2" inches of loose refractory is put in the gap evenly all round and then firmly rammed down until compacted, then the next 2" is dropped in etc.

The first former you will come to is the tap hole, once you have reached this level drop refractory in from the top and add more from the outside of the former to obtain a level where the plastic pipe sits nicely in the middle of the former and firmly against the inner former, make sure there are no gaps under this pipe, the refractory mix doesn't "shuffle" around very well to fill voids.

Now your assistant can hold the pipe in position while you drop more refractory down the gap and ram it up very firmly. Work just the area around the tap hole for now until you have a couple of inches of lining above the former, tapering away at the sides for 4"-5". The hand you are not ramming with can stop the refractory from coming out of the front. Once you have a couple of inches firmly rammed over the height of the former, your assistant can pack and ram from the front, using the water spray sparingly!! to help get a reasonable finish around the pipe, don't try for perfection finishing this stuff off, it doesn't happen, get it reasonable and leave it!

While this is going on you carry on packing and ramming from the top to level things up. When you get to the height of the other former do the same thing again. The first hole to do is the inlet port as it's lowest, make sure you have a firm bed with no voids for the tube to sit on and the clearance around it is even. Push it firmly onto the inner former without pushing it out of place and pack refractory around it, leaving clearance to the side for the pre-heater pipe to go in. Now do the same for the plastic pipe for the pre-heater. This is where your assistant earns their pay; their two hands can hold the pipes in place while you build up the refractory. Once again concentrate on the area surrounding the pipes until a couple of inches of refractory is firmly rammed to hold the pipes in place, then you level off the rest of the refractory layer while your assistant makes good on the outside, make sure that the refractory put in from the outside is firmly rammed in place. I found the best way to finish off these outside patches is to poke it about and smooth it with my finger tips with it slightly moistened with the water spray.

Now just carry on until you reach the top, don't rush things or try to build up too much refractory in one layer, it really does need small amounts packed really firmly to get a good job.

Once you reach the top, go slightly over and "tamp down" with the straight edge, then sliding the straight edge from side to side using the outer skin as a height guide do the best job you can to finish off, a little water from the spray will help but don't get it too wet!

The same thing for the lid, a couple of inches at a time and rammed firmly. A trowel will help to get a finish on the top, once again, it won't be perfect.

As you get near the top of the furnace and the lid you need to try to judge how much refractory to mix up so none is wasted. After mixing the first bag to the manufacturer's specifications, you should be able to judge by appearance and feel to mix up a part bag. Make sure you mix up enough to make some plinths, at least two, but these do need replacing from time to time so within reason as many as you like. Ram them up a little at a time like the lining.

Wait until the next day and generously spray water over any exposed refractory, then cover as much as you can with plastic to seal the moisture in. Leave this to cure now for 4-5 days, the longer the better.

Once it's cured, the inner former can be removed, drop the bottom to allow the hardboard discs to be turned vertical and pulled out. Pull the inside edge of the metal sheet so the former gets smaller and comes out.

Try giving the plastic pipes a twist with a pair of grips, if they don't want to shift leave them until the furnace is lit to burn out. Using a screwdriver or something similar, collapse the food tin and remove it from the lid.

How well did you do with the ramming up? There shouldn't be any nasty voids, any small defects will have to be ignored, the refractory will repair but not really small repairs, I will explain later in the book how to do repairs.

Now we need to make the lid opening foot pedal. I used a small channel section for the see saw bit, but you can use angle iron or anything else you have around that's suitable. Cut 2-3" lengths of 1,1/2" angle iron which will be welded to the bottom plate with the correct gap between them for the pedal to move freely but with minimum side clearance.

If you carefully turn the furnace upside-down, put a piece of cardboard down first to protect the lining, and clamp the bottom plate in position, you can now lay out the position of the pedal. It needs to go under the tube for the bar to rest on it and stick out enough on the opposite side for your foot to be able to operate it. It will need around 1,1/2"-2" total see saw movement, I used a 6mm. bolt with two nuts locked together for the pivot pin. The pictures should explain all.

Turn the furnace up the right way and put the bar in the tube, this bar needs to be higher than the lid height by several inches, if your bar is too long leave cutting it until the arms are welded on.

I didn't have a piece of bar long enough but I had 2 pieces that were half the length, so I used these which means when I lift the lid off for any reason I only lift half of the bar, the other half stays in the tube. This is easier to lift the lid off (something I don't do often), but a pain when the bottom is dropped as the loose half of the bar drops down as well.

The weight of the bar will cause the see saw to drop down that end, put a foot on the foot pedal side to raise the bar as high as it will go and mark the bar about a 3/8" above the top of the tube, this should be the height to fit the bolt which will slide up the slot in the back of the tube and hold the lid open when it's swung out of line with the slot. Line your mark up with the slot and push the pedal up and down to see if you have enough movement of the bar to lift the mark around 1,1/2" (see p.29/31).

If all looks well, drill a hole in the bar at this height and tap a thread to suit the bolt you will be using (6mm. 1/4"?). Tighten the bolt hard to the bottom of its hole, and then cut the head off leaving enough thread to sit on the top of the tube but not enough to catch the side of the furnace.

If you don't have a suitable tap this size you can drill a hole as close to the size of a suitable bolt or piece of round steel bar you might have so it will be a close fit in the hole. Then "bruise" the section which will be going in the hole by hitting it with a hammer on something solid like a vice to distort it, don't over do it! Then either squeeze it into the hole by winding it in with the jaws of the vice, or tapping it in with a hammer.

A piece of flat plate is welded vertically to the bar to strengthen the joint where the arms are welded on, once again use whatever you have lying around that will do the job, this will be in line with the bolt and slot.

Two pieces of flat bar are drilled and bolted to the lid by the 8mm. bolts, put a washer either side of the flat bar then the nut. Although these are done up tight they will allow the lid to align itself on the furnace. Build the arms so they can be welded to the inside of these flat bars.

I used square tubing for the arms as it was the first thing to come to hand, but round will work fine too. The arms are a couple of inches above the lid; I used some wood as spacers to hold the tube above the lid to get everything in position, cut to shape the ends where the arms will join the bar

The piece of flat plate welded to the bar for the arms to attach to

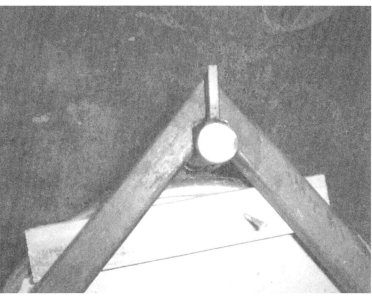

Cutting the arms to fit before welding them in place, wooden blocks are used to space up from the lid

A view of the arms and handle

How the arms connect to the lid

and the welded plate then welded it all together, the round tube for the handle just seemed more "handle like" than the square.

I think the pictures show how everything goes together a lot better than I can explain it.

Make a spout around 3"long from a piece of steel sheet, by bending it into a curve. It will be welded to the furnace at a slight downwards angle so cut the end to be welded to suit and then tack it to the skin under the tap hole as close as you can to the refractory. The spout will have clay based lining on it to stop the metal from sticking. To help this stay in place drill a hole in the middle of the spout. I used a 4BA. (3/16" or there abouts) bolt for this. Put a washer on the bolt and then tighten a nut to this washer, put this assembly through the hole in the spout from the top and secure with another washer and nut. When the clay mix is built up on the spout to the height of the tap hole, the clay will mould itself around the washer and bolt and will be reasonably held in place when the clay is dry; this is changed regularly when different metals are tapped.

Insulation

I had problems with the insulation and what I am using now is a compromise, but one I'm happy to stick with. I'll explain the problems and leave the final choice up to you.

Trying to keep the costs down I first used expanded pearlite as the insulation, this is obtainable from garden suppliers for putting in soil and is pretty cheap. It looks a bit like beads of polystyrene.

I poured this between the skin and the outer former, prodded it about to "settle" it in place and then capped this off with a mixture of fire clay and fine sand (1 part clay to 2 parts sand with the minimum water to mix it).

Although the skin of the furnace became very hot during use, compared to my previous furnace which had 2,1/2" of refractory and no insulation there was a great improvement. I was very impressed until after a very hot trial when I added around 15% diesel to the oil, the capping on the insulation broke up and fell into the gap where the insulation should have been! Over a third of it had disappeared; I assume that the outer surface of the lining exceeded the vaporising temperature of the pearlite and any close enough to the lining vaporised. The temperature inside the furnace exceeded the 1700deg.C. refractory as this melted (even oil alone can exceed this temperature, care is needed)

I would guess that a good operating temperature for melting cast iron (all temperatures in this book are guess work based on what melted at the time) is somewhere in excess of 1500deg.C.(2732deg.F.) If the insulation is efficient, the lining will build up to this, or close to this temperature regardless how thick it is as the heat won't dissipate away.

I didn't even start to look around for insulation capable of operating in excess of these sorts of temperatures it will be available, but I doubt in my budget! Also I was concerned with the damage I had done, not only to the lining, but also I had all but destroyed the clay/graphite crucible inside at the time. (I should point out I was melting steel at the time, I did it, but won't bother again at this price! An accurate form of measuring the temperature is needed to allow things get hot enough, but not so hot as to destroy everything).

Another problem also became apparent, the outer former being steel was starting to look sorry for itself; steel "flakes" when it gets hot enough making the steel sheet thinner with each flake that comes off. My refractory had developed a hairline crack, this will cause no problems at all, my last furnace cracked early in its life and is still going strong 3 years later because the lining is contained inside the skin. But if the outer former wastes away on this one the lining is unsupported.

I decided that the heat being produced by this furnace was enough that I could afford to waste some. By allowing some heat to "bleed" out of the lining through the insulation to atmosphere, the lining would have a larger safety margin before it melted and the insulation wouldn't be subjected to so high a heat. However as the crucible is sitting in the middle of the furnace, it would still build up enough heat from the combustion to melt the metal. If the fuel was costing money this would be a problem, however since a small increase in the fuel consumption and melting times wasn't costing me any more, who cares!

The insulation I've used is still better than no insulation at all, and during testing, melting the lining yet again! And running for a number of hours at these excessive temperatures, the insulation has held up well. I should point out that it's only the surface of the lining that starts melting not the whole mass, so although my lining has a black, glazed surface and a number of solidified "drips" all over, and is a little thinner than when it was built, basically it's still sound. I have had to use the masonry drill, rotary file trick to clear out the pre-heater hole as the molten lining partially blocked this up.

I have gone into some detail with this overheating problem to allow you to learn from my mistakes and keep your lining in better shape than mine! You need to learn how to use this furnace, build the heat up gradually until you are familiar with using it!

Vermiculite is the answer for a cheap workable insulation, this is also available from garden suppliers but I don't recommend you buy it from there as it is supplied in very small pieces, and at a high price.

It is also used as insulation in the building industry who use much larger pieces and quantities so it works out cheaper. I bought it in a 100ltr. bag from a plumbers merchants, try builders merchants as well.

The larger pieces compress well, this more solid mass allows some heat to dissipate through it and is very firm giving a lot of support to the lining, it's surprising how much you can compress into the area we are putting it.

Treat it much like ramming up the lining, pour in a few inches and ram down quite firmly. Stop about an inch or so from the top so we cap it off with something more solid.

Some thought needs to be given to a seal between the furnace body and the lid, due to the large amount of fuel vapour we feed into the combustion chamber and the pressure created by the combustion, this vapour will seep out from under the lid as smoke. It doesn't make a lot of difference to the efficiency, but isn't very sociable!

I bought a reel of 12mm. asbestos rope for next to nothing, however if you look at plumbers merchants who deal with central heating boilers, shops who sell solid fuel stoves, kiln suppliers and any others who are likely to deal in high temperature seals, you should be able to find some ceramic fibre rope to do the job. An alternative is to use a woven strip of ceramic fibre or some thin ceramic fibre blanket cut into 1" wide strips.

If you are using rope, a groove needs to be cut in the insulation capping before it dries, a little under half the diameter of the rope so the seal can sit in it, use a tea spoon or something similar to cut the groove.

If you are using a flat strip, level the capping of as best you can and use fire cement, which can be bought from hardware shops, building suppliers etc, in pre-mixed tubs as an adhesive to stick the seal to the capping.

I capped my insulation off with a mix of 1/3 fire clay to 2/3 fine sand (soft sand for bricklaying is fine). This is mixed dry then the bare minimum of water is added to bind it together, it must be left for a day or so for the clay to absorb the water properly before use. It must be very dry to minimise the shrinkage as it dries, only just wet enough for it to be rammed in place. If you can only just squeeze it into a ball with a lot of effort it should be about right, if it squeezes into a ball easily, then dry it out a bit.

If you haven't managed to find fire clay, try using fire cement with around 1/3-1/2 sand added to it, you must allow it to air dry for a few days before getting it hot, it's not really up to the temperatures we're using so if you come across any alternatives, use them.

Ram the clay on top of the insulation, or pack the cement in and level off. Cut the groove if you are using a rope seal and leave to dry.

Once this is dry, carefully turn the furnace over and pack the insulation under the inlet port/pre-heater area and fit the closing plate.

If you are using a strip seal, mix about 50/50 fire cement and water to make the adhesive. Dampen the top of the capping with water and smear the adhesive on. Place the seal on the cement, if it doesn't like to curve easily, cut it into short lengths and put it on in pieces (try the strip before putting the adhesive on to work out the best approach), then put several sheets of news paper between the seal and the lid, shut the lid and push it down firmly to bed the seal down evenly. After a day or so carefully lift the lid and allow the air to get to the cement to dry, any news paper stuck in place will burn off once the furnace is lit.

I just lay my rope in the groove and push the cut ends together to close the gap (cut it about 1/2" too long), if you have a problem with the rope not staying in place, use a few dabs of fire cement adhesive.

This seal will need to be replaced from time to time so find a good, cheap source for it. The idea behind using a soft-ish capping is that it is easy to replace if it gets damaged or gets a build up of old seal stuck to it.

To prolong the life of the seal, never move the furnace with the lid sitting on it, either remove the lid or I put some wood strips on top of the refractory that hold the lid off of the seal when it's not in use, also try to lift the lid straight up when opening or closing it.

To "make up" the bottom, clamp the steel plate in place and ram up (you should be good at this by now!) with the same clay/sand mix used for the capping until you reach the bottom of the tap hole. This depth will vary depending on how your tap hole went in. If you didn't find any fire clay, you really need some clay in this mix. Do you have any clay in the soil around where you live? I have used this to line furnaces, and often use it in the bottom of mine mixed with sand. It melts at the temperatures of iron melting, but then so does the fireclay/sand mix. If it's thick enough you'll be alright and it's cheap!

Cupola furnaces use sand alone, however I find the hard bottom with the clay reflects the heat better, and also you'll find you get oil leaks from the bottom of the furnace, although you sometimes get a little leaking using clay/sand as well. This bottom needs to be hard, but easy to break out when needed.

Never drop the bottom when the furnace is hot, the thermal shock can crack the lining. Wait until its cold then drop the steel plate and knock the bottom out with a steel bar, starting from the middle and work your way out.

I use the same 1part clay/2parts sand mix for the "bod" to plug the tap hole. I have a bucket with a lid filled with the mix ready to go all the time; it lasts indefinitely if kept slightly moist.

The pipe work

If you are making your furnace to the same dimensions as mine then stick with all the sizes I give in this book, if you are adapting your existing furnace or building a different size from new to use my oil burning set up then there is something to bear in mind. There is a direct relationship between the internal volume of the furnace, the air supply and the tapered cone that creates the venturi effect to draw the oil out of the oil pipe and into the air stream. A smaller volume combustion chamber will need less air fed to it for the amount of fire it can hold. If this smaller amount of air is not enough to create enough venturi to draw the oil out, then the cone outlet hole must be smaller to increase the air velocity over the oil pipe.

Once the pipe work is completed, it can be set up outside of the furnace and the air and oil turned on (use a bucket to catch the oil). If the oil dribbles out of the pipe and doesn't get picked up in the air stream, then more air is needed or the outlet hole in the taper should be reduced. If most of the oil gets picked up and a little dribbles, this is probably fine as when the furnace is lit the pre-heater will be thinning the oil down. If a larger pipe is needed to increase the air volume for a larger unit, then the same test can be used to get the taper/air flow right. Once the furnace is lit, if the exhaust flame is too big to get it burning right then less air and a smaller outlet hole is needed.

I had a length of 2" dia. thin walled stainless steel pipe in the scrap box, 2" is the minimum diameter you should use and up to 2,1/2" would be fine, although you will have to play around with the hole size at the taper.

Thin walled pipe is needed as we want heat to pass from outside the pipe to heat up the air inside the pipe as efficiently as possible, I have used car exhaust pipe in the past, but the stainless pipe I am using now is thinner.

Stainless pipe is going to be longer lasting, but the air passing through the pipe picks up most of the heat, I have managed to get the heat shield red hot in places but the air pipe never gets this hot, so it won't burn away too quickly if you use mild steel.

If you are using a hoover blower with the plastic T air control, the air pipe wants to be about 20" long. If you are using a metal pipe/fan set up the pipe can be shorter. You will have to see what works out about right for your set up based on the pictures of mine.

Cut one end of the pipe as straight as you can, by wrapping a sheet of paper around the pipe, as long as where it overlaps the edge of the paper lies directly over the layer underneath, you can mark the tube with a felt tip pen

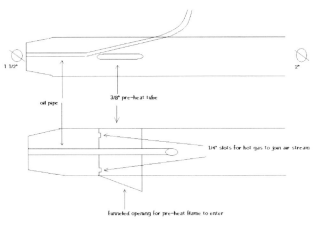

The lay out of the air pipe

The six segments with a slight twist ready to taper down

The finished taper

Using paper to mark the position for the 3/8" slot

from this edge to get a straight line. This can then be cut with a hacksaw, once the saw has cut through the first bit, keep turning the pipe so you only cut one side of the pipe (usually the side nearest you), this way you can follow the line and get a straight cut.

Using the paper again, mark another circle 1,1/2" from the straight edge, now cut the top of the tube into six equal segments down to this line. Using a pair of pliers, give each of these segments a slight twist the same way. Put this end of the pipe on a hard surface like a vice and tap the ends of the segments down to form a taper with an 1,1/2" hole. By putting a piece of 1" round bar or solid tube in the vice, the taper can be slipped over this bar and the segments "dressed" by tapping with a hammer on the outside against the bar to make a reasonably finished cone.

Slip this pipe into the inlet port of the furnace so that it stops 1/2" short of entering the combustion chamber, if the cone is too far into the chamber too much heat reflects back and vaporises the oil in the pipe causing blockages, with it finishing short of the chamber the air flowing through the pipe keeps it cooler. When the air pipe is finished, a jubilee clip (hose clamp) can be put around the pipe so that it acts as a stop when it hits the inlet port tube so the pipe is always the correct distance in.

We need to mark where the flame from the pre-heater hole will hit the air pipe. Place something in the pre-heater hole (I used a rolled up piece of paper) and mark the centre of where it touches the air pipe, this will be the centre of the slot we need to make in the pipe.

Digging in the stainless scrap box again I found a piece of 1,1/4" dia. thin walled pipe from a carpet cleaner wand (once again mild steel will be ok.). I cut a 3" length of this off and squeezed it in a vice until it was 3/8" thick. Place this over the mark we made on the air pipe so the mark will be at the centre of a 3/8" high slot, and mark where the ends of this flattened pipe are on the air pipe. This flattened pipe needs to go in one side of the air pipe, through the middle, and out the other side. If you don't have access to a mill, to make the slots on each side of the pipe, drill a 3/8" hole at either end of where the slots are going and cut between them with an angle grinder.

Before we fit this flattened pipe, cut or file two 1/4" slots in the edge of the pipe which will be nearest the furnace. These will be inside the air pipe to allow hot gases from the flame passing through this pipe to join the air stream flowing into the furnace.

One end of the flattened pipe is welded flush to the air pipe, the other end will stick out 1", weld it to the air pipe where it comes out, then cut this end at an angle which runs from the air pipe (edge nearest the furnace) to the

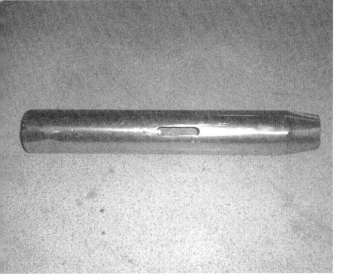

The 3/8" slot for the flattened pipe

The hole for the oil pipe, with the flattened pipe welded in place on the right, and a smaller dia.extension pipe welded on to the left, this is for a hoover blower

The air pipe, oil pipe and valve complete

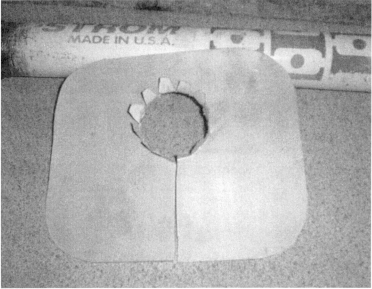

Part 1 of the heat shield with the points
Cut off from the segments

end of this inch overhang, then widen this part out with pliers to form a sloping funnel to guide the flame from the pre-heater hole through the flattened pipe. Around half the heat from the pre-heater hole will go through this pipe; the other half goes around the outside of the air pipe to heat it up.

When the air pipe is in position, this flattened pipe runs horizontally across it with the funnel end to the pre-heater hole. About three inches back towards the blower, drill a hole for the oil pipe in the top of the air pipe. The oil pipe is 10mm. dia. I drilled an 11mm. hole, then put a piece of metal bar in this hole and bent the back of this hole down and the front of it up as far as I could get it to go, so that when the pipe is fed in the hole it would run towards the taper.

The copper oil pipe needs a slight bend to pass over the flattened pipe, and a gentle curve as it comes out of the air pipe, the picture should show the idea. It needs to finish flush with the end of the taper and in the middle of the outlet hole, and have about 2" sticking out of the top of air pipe. I brazed the joint where the oil pipe passes through the air pipe as I used stainless steel pipe, if you are using mild steel for the air pipe it would be alright to soft solder it.

If you want to fit a propane pipe, solder/braze/weld a stub of suitable pipe somewhere around where the oil pipe enters the air pipe. I used a piece of small iron gas pipe with a cap to close it off. I made a brass gas jet that slips into this gas pipe, about three inches long with a 1/8" hole through it, which connects to the rubber gas pipe. I only use propane to start the furnace off. You will also need a high volume, adjustable gas regulator.

If for any reason you want to use propane or natural gas as a main fuel, I would look at running it through the oil pipe, I will leave any experiments of this nature to you as I can't afford to do it myself, and am not qualified to advise you on using gas!

The heat shield is made from thin steel sheet; the pictures show how to cut them out. On part 1 bend the segments to form the 2"dia. hole and trim the pointed ends off leaving enough for a jubilee clip to go over to secure it to the pipe. Bend it open enough to go over the air pipe and jubilee clip (hose clamp) it securely. Over-lap the outside edge of the slit by a 1/4" and tack weld together so it cups towards the furnace.

Bend part 2 to go over the air pipe and tack weld to part 1, have only about 3/8"- 1/2" clearance between the air pipe where the flattened pipe goes through and the heat shield. This is to allow a little pressure to build up in this area to help the flow of hot gasses pass into the air stream through the slots we cut in the flattened pipe.

Part 1 of the heat shield clipped in position

Part 2 tack welded in place, note the oil pipe level with the front of the taper

The "Hoover in a Box" assembly ready for use

My home cast blower

The fuel valve is described in the casting section of the book, it needs 6mm. and 10mm. threads tapped into it. If these present a problem then a small G clamp can be used to control the fuel flow. A ball lever valve is best to use as an on/off valve as it's only 1/4 of a turn to shut things off.

The picture shows the oil tank and feed set up. A steel oil tank is best, with a 15mm. (1/2") pipe soldered/brazed to the bottom of it for the oil pipe. However I didn't have a 10litre metal can, so I used a plastic one with a 15mm. tank fitting in the bottom of it to connect it to the copper pipe. It has the advantage of being easy to see the oil level, however there is a safety issue with the chance of it melting, your choice!

Repairs

Small defects/repairs are difficult as the refractory is "lumpy" and not very sticky. Fire cement is smooth and easy to use, but can't withstand the higher temperatures! The suppliers I use for my refractory do stock "smooth" high temperature mixes, however the cost and the cost of delivery make it impractical.

I have used, and will keep using fire cement for small bits and pieces and replace it regularly as it falls out. A fire clay mix can be used sometimes in the same way. But we are talking about a furnace here! If it gets a little beaten up but still works, then don't get too worried!

Whatever is used for repairs the process is the same. Remove any loose or dusty debris (a large repair is easier than a small one so don't be too cautious!), and then thoroughly dampen down the area with water, go beyond the repair site. I sift using a tea strainer, some dry refractory cement to get a fine-ish powder which I mix with water to about the consistency of single cream; both fire cement and clay need this also. Paint this over the wet repair area and immediately apply the material for the repair before the lining absorbs the moisture. Make the repair mix a little wetter than was used for the lining and work it into the repair as best you can.

If necessary use a former to keep holes open e.g. the tap hole, or inlet port, and allow a number of days to dry (a lot of the time, fire clay is best heated up straight away and the heat built up slowly).

I've had to do some repair work on this furnace due to a failed attempt to melt and pour 35lbs. of cast iron directly in the furnace, I ended up chipping out the tap hole back to the metal former and re-casting it, so far I've had no problems at all with the repair.

The inside of my lining has a black, glazed finish due to melting it on many occasions, this must be chipped or ground off back to plain cement before attempting any repairs to it.

The hoover in a box set up melting cast iron, burning used cooking oil

Using the furnace

Safety!! Water and molten metal don't mix, since any molten metal is many times hotter than the boiling point of water, as soon as the two come in contact the water turns into super heated steam which expands very rapidly blowing the molten metal everywhere, it's much like an explosion, keep the two apart! Also any metal spilt on concrete will cause it to blow apart. I use metal trays filled with **_DRY_** sand to put crucibles down on or under the tap hole, a better idea would be to have a bed of dry sand all around where you are working so any spills would be safe.

I must confess, I'm not the most safety conscious person around, but when working with molten metal you must think safety! When I first started melting aluminium, my only concessions to safety were a pair of stout leather boots and a pair of welding gauntlets.

With this furnace that's not good enough! I now use a leather welder's apron, welding goggles, and leather arm covers with the boots and gauntlets, a full face mask would be better than the goggles. The brightness of the exhaust flame and a crucible of molten cast iron mean some form of tinted lenses should be worn, even sun glasses would be better than nothing. Your safety is **_YOUR_** concern, be sensible!

Don't even think of melting cast iron until you have a reasonable amount of experience with aluminium, you need to develop a good working practice around the furnace, also if your moulding sand is too wet the reaction you get from aluminium is a lot calmer than that with cast iron! The heat radiating out from both the furnace and the crucible is something else.

It's best to start things off with wood to give the lining a chance to dry out fully, feel free to melt some aluminium during this stage. With wood you will get a lot of embers and sparks blowing out from the exhaust and pre-heater holes, make sure nothing around you will catch fire

Set everything up with the oil ready to go and the oil tank upwind of the furnace, have a good pile of dry wood cut to a size that will go in the furnace. The idea is to have as much wood under the crucible as possible, and to drop pieces vertically down along side of the crucible as well, but leave enough gaps for the air to circulate. A two pint metal pot works well, but if the metal isn't hot enough to pour as the wood burns down, the pot can be lifted out and more wood added before replacing the pot. Always turn the blower off before opening the lid!

Light some kindling in the furnace however you would light a fire, and close the lid. Turn the blower on to the gentlest blow, and allow this wood to catch. Turn the blower off and open the lid, there will be some smoke as not enough air will get to the wood for it to burn. Quickly fill your furnace with wood (and the crucible), close the lid and turn the blower on. Allow the wood to catch gradually increasing the air as it does. When all the wood is burning the exhaust flame should be bright and full and not wavering about too much. Not enough air and it will smoke and waver about, too much air and the flame will get thin in appearance and dimmer, this will create less heat.

You can only learn the correct amount of air by playing around, the fuel costs nothing, the scrap metal costs nothing, take some time to get the hang of it. I won't say anything about pouring the metal as it is covered later in the book.

Don't turn the oil on until the furnace has stopped steaming; allow the wood to burn down quite low. Any time you use wood to start the oil, you won't get a "proper" oil burning flame until all the wood is burnt away. You will learn how much wood is enough to "start up" without overdoing it, but does it really matter? It's only a few minuets of a lower heat out put. Actually the same can be said of starting up with propane; until the gas is turned off you have two fuels competing for the same air supply.

If you are using a G clamp to regulate the oil it should be clamped tight, with the fuel valve have the needle valve fully in. Fully turn on the ball lever valve and start opening the oil regulating valve, remember the oil is thick! It will take time to flow down the pipe into the furnace. Should for some reason a huge flame comes out of the furnace, turn off the ball lever valve. However it's unlikely this will happen, with the cast fuel valve, once it's been used once it will be more or less at the right setting all the time, only needing a little adjustment depending on how hot you want to work, or how hot the furnace is at the time.

The air will need turning up as the oil flows into the furnace as you will have it set for the wood burning and now you are adding more fuel. You will tend to have a bigger exhaust flame burning the two fuels, this will come down as the wood burns away (or the gas is turned off). Something around 12"- 15" is normal with either wood or propane when oil is added.

On this occasion we are using the wood to dry the furnace out, if you are using the wood just to start the oil then put enough wood in to get a good blaze, after four or five minutes turn the oil on. Once you are used to the furnace and if the weather is warm, this time will be shorter.

If you start up with propane, use a 24" long piece of wire with one end bent over trapping a small scrap of folded cotton rag under the bend. Dip this in paraffin or bbq. lighting fluid. Open the lid and turn the blower on very low, light the scrap of rag and put it in the furnace in line with the inlet and slowly turn the propane on. As it ignites, adjust the gas to a quiet wavering flame then turn the air up until it roars, turn the gas up some more, then more air. Now shut the lid, if the fuel/air flow is too gentle the gas can ignite in the air pipe when the lid is shut. If this happens turn the gas off, open the lid and start again.

When you are burning oil or gas, the air is the first thing on and the last off. The air must be on to blow the fuel into the combustion chamber, otherwise it can burn in the pipe or leak everywhere. Turn the air off first and the same thing can happen, as well as having fuel present and no air to burn it which will cause smoke, it's possible for it to suddenly ignite as the lid is opened and the air reaches it.

Once it has been running on propane for a couple of minutes turn the oil on and increase the air flow. After about a minute or so the gas can be turned off. Remove the gas pipe and move the gas bottle from the area.

Although both the wood and the propane will start the pre –heating going, it's not until the furnace is full of burning oil that it really starts working, as the wood burns away and the furnace starts heating up we can gradually increase the oil flow and adjust the air flow to suit until it's running as hot as we want it.

By glancing through the exhaust hole you might just see the crucible through the fire, if you can see it quite clearly there is not enough fuel burning, or too much air. You might get a very small amount of smoke from the exhaust, but if it is a lot there is too much fuel or not enough air.

The air/fuel mix is easy to determine, a very orange flame with smoke is too rich (too much oil), whereas a thin yellow flame is too lean. If you can see into the furnace clearly, or the pre-heater flame is very small, turn up the fuel and air. If you have a large exhaust flame you are burning too much, somewhere around 8"-10" inches high is about right, but this isn't an exact science!

Because the metal is being melted inside the combustion chamber, any change to this metal, i.e. topping up with cold metal or placing a new charge of metal in the furnace after a pour, will alter the overall temperature in the combustion chamber. This will mean that it is sometimes necessary to slightly alter the fuel flow until this new metal heats up.

When the furnace is turned off during a melt, to re-light, simply turn the air back on followed by the fuel and it should re-light. If you are dealing in a low heat at the time and the furnace has been turned off for a while, you might find it will smoke but not ignite, turn the fuel off and use the lighting wire or a piece of burning wood etc. to ignite the vaporised fuel as you turn it back on, although it's quite rare for the furnace temperature to drop this low during use.

When you are melting cast iron the exhaust flame will start getting smaller and brighter as the iron starts to melt, this is a good indication of the progress of the melt, it is due to the furnace interior getting so hot that combustion is very good, with the better insulation this happened a lot quicker, before the iron started to melt.

When you try your first cast iron melt you should be used to the furnace from melting aluminium, just leave the heat to build up, and if necessary turn up the heat gradually until the iron melts, remember it takes time for the heat to penetrate through the crucible and to be absorbed by the iron, allow around 30mins for an A4 crucible to melt the first charge of iron (from a cold start), and between 45-60 mins. for an A8 crucible. But if it takes longer on your first attempts good! You are taking things easy and not destroying anything.

These times are a guide only and refer to the starting quantity of metal that you were able to load into the crucible, and melted does not mean pouring temperature but fluid enough to skim the slag.

Around 10-15 mins. after start up, the fuel/air can be turned down if a lower heat is needed. As long as there is enough air passing through the air pipe to draw the fuel out it will run fine.

Your blower settings will be the best guide for re-setting the furnace to a temperature for a particular type of job. The oil will vary with every batch you get; also the moisture in the air and the air temperature will make a difference, so every time you use the furnace it will be slightly different.

If with your blower running flat out and providing your oil/air mix is about right you can't melt cast iron, then you will probably need a more powerful blower, however, if you blow too much air/fuel in, you can cool everything down and get the same result, another reason for starting low and building up!

One annoying thing with this furnace, when the fuel is turned off there is a small amount of oil in the lower part of the oil pipe which vaporises as soon as the airflow stops. As there is not enough air for it to burn properly you will get an orange flame with a gout of smoke. Also with engine oil it is possible for the very end of the oil pipe to get blocked with carbon if this happens a few times. This doesn't happen so badly at lower temperatures but is a pain when melting iron, the bleed screw on the fuel valve helps, by opening this when the fuel is turned off, then as the flames stop, turn off the blower, most of the oil is out of the pipe but not all!

I now turn off the oil, then turn off the air and remove the air pipe from the furnace. A drip tray is needed under the inlet port to catch any oil drips when you pull out the air pipe, a few drips still vaporise and you will always get some smoke. When you replace the air pipe and turn the air then the fuel back on, the first oil will enter this hot pipe and vaporise too quickly, this will cause a larger exhaust flame than normal, just stand back, it will settle down in a few seconds. This orange flame creates soot which gets the furnace (and you if you get too close) sooty, a damp cloth wipes most of it off (when the furnace is cold), although I prefer a hot shower!

If I am just checking the progress of a melt, most of the time I turn the fuel off and leave the blower going, a quick "poke" with a long poker through the exhaust hole reveals a lot!, I have also dropped top up metal through the exhaust hole or given the pot a stir. It's not good to leave the blower running with no fuel for more than a very short time as the "superheated air" has a cutting or scouring effect and can damage the crucible and lining. Also the metal deteriorates as it absorbs an excess of oxygen.

Burning engine oil, you will get carbon deposits build up on the inside of the furnace directly opposite where the air pipe blows in; these will cause the air/fuel mix to deflect upwards. Every time I open the lid to skim the slag I scrape these deposits off with the slagging tool, they come off easy enough. You will find something that looks like sand in and around the furnace when burning engine oil. I puzzled over this for a while, and decided it was the burnt remains of the carbon. These faults are a nuisance, not a major problem.

Unfortunately cast iron doesn't melt directly in the furnace; it must always be in a crucible. There are several reasons for this; firstly heat or hot air rises so the bottom of the furnace is the coldest part. When the air/fuel mix enters the furnace there's enough to fill the furnace with fire, so it needs to spread out before it all ignites, therefore there is a colder area immediately in front of the inlet, this deflects off the furnace side, some downwards, which slightly cools the bottom in this area. The bottom is un-insulated (I have tried insulation below the clay as it was rammed up) so there is a great heat loss with the metal in contact with the furnace sides and bottom. However any pieces of iron you drop in will melt quite well up to a certain amount, and then the heat can't penetrate through the mass enough to overcome the losses.

With any molten metal, it oxidises when comes in contact with the atmosphere, the metal deteriorates to form more slag. When the metal is in a crucible a lid can be used, or the slag itself creates a barrier from the atmosphere. The percentage of metal that deteriorates in the crucible to form this slag is quite small given the depth in the crucible, therefore when the metal is stirred before pouring the overall quality of the metal should still be good.

Whereas an inch deep pool of metal 10" dia. is allowing a far greater percentage of the metal to be in contact with the atmosphere and the metal deteriorates to a much greater extent. The longer it sits in the bottom of the furnace trying to get it hot enough to pour, the more slag you will create and the less metal, which would be of very poor quality if you did pour it! Meanwhile the heat will build up so much in the rest of the furnace the lining suffers very badly. It simply isn't worth it, I speak from experience!

You might think, it works fine in a cupola type furnace so why not this one? (this was my thinking as well to start with). However cupola's "super heat" the metal as it drops down through the coals, so when it arrives at the bottom it is well above the pouring temperature and slowly cooling down. Also the coals and ash the metal is surrounded by reduce the contact with the atmosphere.

I have found no problem getting copper or colder melting metals nice and hot in the bottom, although you must get them out as quickly as possible to keep the quality up.

The direct melting ability of this furnace really comes into its own for reducing large pieces of scrap metal that are too big to go into a crucible, or even into the furnace itself, down to ingots.

An aluminium car wheel can be placed on top of the pre-heated furnace with the lid open, a few bricks help to form a partial lid. Fire up the furnace again and let the well fill up to the level of the inlet port. Now you can either tap from the spout, or ladle out from the top into ingot moulds. Long thin pieces of metal can be fed in through the exhaust hole, it is a very quick and cheap way to "stock melt" that pile of scrap that is taking up so much space into a nice stack of ready to use, clean ingots. Since the bottom can be dropped and replaced so easily and cheaply, a new bottom can be put in to stop any contamination from different metals.

To plug the tap hole we use the same 1/3 clay 2/3 sand mix to make a "bod" simply roll some into a ball a bit bigger than the hole dia. If the spout lining is fresh, then fire the furnace up with no bod so the flame coming out of the tap hole dries it. Then turn the furnace off and drop the bod on the spout and push into the tap hole with either a piece of wood or the bod stick I describe later. If you are tapping metal and need to stop the flow, then work the bod onto the bod stick and push it into the hole, it will dry almost immediately, hold it in position for 5-10 seconds. To tap the bod, simply knock a metal bar through the bod with a ladle under the spout; don't push the bar in too much, just enough to create a clear hole for the metal to flow.

By using the food can/refractory plinths, the crucible is held above the coldest part of the furnace right in the hottest area, if you wanted to melt a huge crucible of aluminium a shorter plinth can be used, a solid piece of steel tube might get you out of trouble.

When you are starting an iron melt, a few pieces of cardboard between the crucible and plinth will leave an ash layer which hopefully will prevent the plinth from sticking to the crucible when you lift it out, it doesn't always work! I always have the lifting tongs for my A4 crucible at hand to put this very hot plinth back into the furnace just in case. The crucible always needs to be returned back to the furnace to cool down slowly, since the furnace bottom might be melted or have an assortment of melted metal on it, the plinth needs to go back for our highly expensive crucible to stand on.

Always try to shut the lid after lifting a crucible out to keep the heat in and to help prevent thermal shock from cracking the lining. When you have finished melting, place a brick over the exhaust hole, and I have a "plug", a piece of pipe with an end welded to it which goes in the inlet port. These will allow the furnace to cool down slowly and prolong its life.

I mention this with caution, the furnace will burn diesel or paraffin, beware of the heat output! And remember to turn the fuel flow right down to start as these will flow a lot faster than waste oil. While it occurred to me to fit a heating element in the fuel tank, both to pre-heat the oil for better normal burning, and to help the flow in cold weather. I dismissed this as too much trouble. I have increased the oil pipes to the diameter described in this book to overcome any flow problems. However if you have a very thick oil combined with a very cold day, thinning down the fuel with diesel or paraffin is an option, be careful, it doesn't take much to send the temperature sky high! This is also an option if you are using a larger crucible for iron and need to boost the heat.

P.S.

This is a P.S. to the end of this section. This is something I have played with for a few years with different furnaces, but haven't come to any firm conclusions about. The air flow creates a vortex within the furnace, this spirals its way up the combustion chamber until it reaches the lid where it now concentrates itself to exit through the central, circular exhaust port.

By altering the exhaust port, we alter the whole flow within the furnace, if the port was off set from the centre the smooth vortex flow would have to end for the gasses to exit. Likewise if the exhaust port was a rectangle rather than round we would get the same effect. The exhaust could even exit from the side of the furnace.

By placing a piece of fire brick partially over the exhaust port we can make some changes to the flow with no structural alterations, and it does make a difference. Firstly, by reducing the size of the hole we create more backpressure which will alter the amount of air entering the furnace and keep some of the heat generated in the combustion chamber longer, however the balance between allowing enough fuel in to generate the heat in the first place, and retaining that which we do generate must be right.

Also we are changing the shape of the port which will alter the flow. Since we will be using different sized crucibles, different fuels and different amounts of heat, I'm not sure there's a permanent change which will be best for all occasions, but I'm sure some experimentation with the exhaust port in general could improve things some of the time.

Making experimental lids from a clay based refractory would be very cheap, and maybe worth while!

Crucibles and metals

Up to this point of the book I have been able to write with some authority. I put this furnace together based on trials and experiments I have carried out, therefore I believe I have a reasonable knowledge of the subject matter.

From here on in with this book this is not the case. Yes I have done a certain amount of casting with different metals, but I consider my knowledge to be very limited. Some of what I am about to write is based on personal experience; some is based on research with information I haven't tried myself yet. Foundry work has been going on for a long time, you will never stop learning. Any information you can pick up will be useful to you but the best way to learn is simply to get out there and do it!

I have given you the tool with which to do the job, now I will try to give you a basic knowledge of casting to get you started.

Zinc alloys, these are used everywhere from the die cast toys we played with to machine parts. It tends to be a dull grey in colour, unless it's freshly cast, and is heavier and harder than aluminium. It is a very useful metal for the home caster and will actually melt on top of a gas cooker in a saucepan (my wife was NOT impressed when I did this in the kitchen!). Since it needs so little heat to melt, burning wood in the furnace works well for it and for small quantities of aluminium. Care is needed not too overheat and vaporise it if you use oil, otherwise treat it like aluminium.

There are two schools of thought with casting aluminium, one says melt it in anything it will melt in, the other says you must use a proper crucible and nothing else will do!

What is a proper crucible? The most common are either clay/graphite, or silica/carbide, this refers to the material they are made from. I have never used a silica/carbide crucible. They are intended for lower temperature metals, I believe the max. operating temperature is around 1450deg.C. and they get eaten away by cast iron slag. They are supposed to be very tough and forgiving in inexperienced hands.

Clay/graphite are suitable for higher temperatures, I can't tell you how much weaker they are as they are the only ones I've used, however I haven't broken one yet. I use Salamander crucibles, they have a good reputation and are ready to use as you get them. Other makes may need running them up to a high temperature to anneal them before use.

I only use "shop bought" crucibles for cast iron, for all other metals I make my own. I made a potters wheel and tried throwing them myself (HA!), probably the less said of this the better!

For melting aluminium, brass, bronze, silver, gold or copper (not that I've had the chance to melt gold or silver, if you have some to spare?) both types of crucible would be fine, and as long as you weren't too clumsy will have a very long life. Cast iron on the other hand limits the life span. The immense heat and scouring effect of the hot blast slowly (can be quickly if you turn the heat up too much!) erode the crucible away, but more of this when we discuss iron.

I have melted aluminium in everything from a baked bean tin (in a low heat furnace only, and only use once) through various saucepans, to cut in half fire extinguishers as well as countless steel and stainless steel pipes with bottoms welded to them, and of course directly in the furnace. Also I now make cast iron crucibles.

The proper crucible only school say, and quite rightly so, that the aluminium picks up contamination from a metal pot and a greater chance of bubbles in its make up than it will have with a proper crucible. This definitely means you will have to tell NASA that if they want those space shuttle parts up to full spec. that they will have to stump up the price of a suitable crucible or go elsewhere!

Let's put things into perspective, what are we making? Probably most of what the home caster makes won't go in the reject (or re-melt!) bin if it has a small defect (or two), and if you do want to take more care over a particular casting it doesn't mean digging deep in your pocket. I prefer to use stainless steel or cast iron rather than mild steel for crucibles, steel "flakes" when very hot and these flakes can break up in the molten metal and get included in the pour. Cast iron and stainless still flake, but much less than steel.

The 1/3 clay, 2/3 sand mix used for the furnace bottom can line the metal crucibles, around 3/8"-1/2" thick works fine and has a few advantages to metal alone. Firstly unlike a clay only crucible (a reference to my misshaped pots), the metal gives it plenty of strength; the lining will prevent the molten metal from picking up any contamination from the metal shell and the thickness of the clay gives more insulation to the pot and a greater "heat store" allowing more "pot time" before the metal starts to chill. Since I mainly use clay I dig locally for this, it costs almost nothing (that's my kind of price), so having different size crucibles and separate crucibles for different types of metal (a very sound idea) is no problem.

Firstly drill a couple of holes in the sides of the pot around 1" from the top to fit a couple of bolts through, heads on the inside, nuts outside. You can always use these to lift the pot with and they prevent the clay falling out when it dries and shrinks. I push the dry-ish clay mix into the pot too thick until the pot is fully lined. Then scrape it out and smooth it down using a spoon to the correct thickness. I then put it in the oven at a very low heat until it's dry, if it has no cracks it can then be used and will fire during use in the furnace. They don't last forever, but who cares! They are easy enough to re-line or to make new and are suitable for any metal with a lower melting temperature than the pot and the clay. Once again, stainless lasts longer than mild steel and is stronger at higher temperatures.

I only use clay/graphite for cast iron, all other metals I will either melt directly in the furnace when suitable or in a lined or unlined metal pot. This makes home casting a very cheap hobby, especially with no fuel costs!

I think of aluminium in three categories, the good, the bad and the ugly! I don't know enough about it to break it down any further. The good consists of pistons and wheels, castings made from these will have a low shrinkage rate, good wear resistance, and will machine to a good finish.

The bad, this is a bit unfair, it's not that bad at all, is normal cast aluminium, cylinder heads, engine blocks, gearboxes etc. If it's been cast and then machined it will do it again, although probably not as finely as "the good".

The ugly is the most common and still has plenty of uses for those "not quite so important jobs" this is extruded aluminium, door frames, ladders, greenhouses, tubing etc. This will shrink quite a bit and you will struggle to machine it to a fine finish, but it still files, drills and taps fine. I tend to use this for all those jobs that don't warrant breaking into my stash of good stuff (despite the fact that none of it costs me money!). The fan and fuel valves described later in this book are good examples, although you can get away making almost anything with it. There is one other group worth mentioning, cooking utensils, good quality pots and pans are a corrosion resistant alloy, a useful thing if you want to make something for outside, boating, camping etc.

Many metals are specialised to suit the purpose they are used for, look for scrap that has been doing the same sort of job you want it for and you won't go far wrong.

Brass and bronze are both alloys of copper, brass with zinc, bronze with tin, but there are many other variations and blends. Having no real interest in casting a nice bronze bust at this time, any bronze work I can foresee now or in the future would be for bushes. I have a growing collection of worn out bushes ready for the job, if they were good to do the job once they can do it again! In theory bronze will melt either in a pot or directly in the furnace

This is my last furnace in action, tapping copper in to a clay lined stainless steel saucepan "ladle". The ball on top of the bucket is the clay "bod" ready to block the tap hole when the ladle is full.

A steel casserole dish lined with clay to make a ladle

¾ pint cast iron crucible moulded from a tea mug

More cast iron crucibles, look like flower pots?

with no real problems and has been melted in cupola furnaces for centuries, but if you're unsure of the make up of the metal treat it like brass.

Brass gives a few problems to home casters mainly because its zinc content will vaporise as soon as it melts. Leave it for a while in the pot and you will end up with a scummy pot of copper, it's not a good idea to direct melt brass! Any dross or slag should be left untouched and the dirtier it is, the more chance the zinc has to stay put as the dross will settle across the top of the molten metal and keep the air out, the zinc can only vaporise when there is air present. Borax can be used as a "chemical lid" as it will melt earlier than the metal and float to the top to form an air tight skin, I have heard of people using crushed glass to melt as a skin but haven't tried it myself (yet). Of course, a crucible lid can be used, but it will have to be lifted to see into the pot letting the air in.

I tend to use plumbing fittings for my brass; they were cast the first time round. I put borax in the pot, enough to form a complete seal over the metal once it has melted (a couple of spoon full), and then it's down to an educated guess as to when the metal is hot enough, any prodding about will allow air to the metal and some of the zinc will burn away. I use a large spoon or skimmer to hold the dross and borax back and pour with it in place. Sometimes I will give it a quick stir first, especially if I see the telltale flash of yellowy white smoke from the pot, you often find some separation of copper. The finished casting will always be darker than the original metal, if you want to you can have a go at topping up the zinc content. Remember you can create any alloys you want, perhaps try to give a metal some properties of another metal, make something softer or harder, burning waste oil it doesn't cost much to try things out!

Copper is pretty straightforward, I haven't done much, I've just heated it up both directly and indirectly (with or without a crucible) in the furnace and poured it.

At this point let's talk about ladles for direct melting; these are very complicated and extremely expensive! Mine might look like old saucepans with stronger handles welded on and a clay/sand lining, but they are far more complicated than this, let me sell you some at a very reasonable (extortionate!) price. Need I say more? Even for aluminium they want to be lined to retain the pre-heating and insulate the metal until it's poured. To pre-heat them, place them above the exhaust flame until the interior is red hot, place them on bricks standing on the furnace lid or make a stand up for them, your arm will get tired if you try to hold them for too long! Make sure you don't restrict the exhaust flow too much. A steel, stainless or iron saucepan with a couple of bolts fitted to help the clay stay in (like the crucibles) and either weld or bolt one or more handles in place capable of handling the combined weight of the pan, clay and metal. I find wider pots allow the heat from the exhaust to heat up the inside better than narrow tall pots.

O.k. cast iron, firstly a clay/graphite crucible is best, if the maximum heat for a silica/carbide crucible is around 1450deg.C. and we're running the furnace at between 1500-1600deg.C. to melt iron, even I can work out this doesn't look good! However as we are using very high temperatures this limits the life of the crucible, run the temperature of the furnace up gently and the longer your crucibles will last. Run the furnace flat out, especially with a smaller crucible, and the heat generated and the force of the blast will erode the crucible away very quickly as well as damaging the lining.

We will do all we can to keep the metal soft, it's more than we can really do to make hard metal soft so we have to start with soft metal. A hacksaw will show you how hard it is, although you will be able to tell a lot by breaking it with a hammer and examining the broken edge, good iron looks very grey with no white or white crystals in it and a very fine grain. As with all metals, look where it came from, if it has been machined before you're in with a chance.

For the cast iron crucibles I make any old stuff will do, a good way to use up sprues and "the ones that went wrong" Every time you re-melt metal, any metal! It looses quality, remember this. With some castings it doesn't matter, with others it does. With pistons for example, since they will go into a crucible "as they are" I would save them for a good job and melt them from their "virgin" state. Car wheels on the other hand will need reducing down to a suitable size, if you want to do this with a hacksaw go ahead but I prefer to reduce them using the furnace.

Cast iron can seem soft but have hard spots in it, as I mentioned earlier, slag inclusion and hardness control are the biggest problems with cast iron!

Let's have a look at what we can do about hardness, firstly pick your iron well, thin sections often are softer then thick and certainly melt down faster for us, remember the faster it reaches pouring temp. the less exposure it has to the atmosphere and the better the quality of the metal and less slag will be produced. A lid will keep a lot of the air out and not piling it too high out of the top of crucible will help.

I have heard of mixing crushed charcoal or coke with the iron and can see some logic to it. Cast iron is an alloy of iron ore and carbon, other things happen when it's smelted but we'll ignore that for now (because I don't know myself!). Some of the contents of the iron that burn out during our pour might well be replaced by picking it up from the high carbon content in the crushed material. Also the air is now restricted in its access to the metal as the coke or charcoal will be filling the gaps between it. Coal will not be so good due to the higher level of impurities in it.

In engineering, the term "normalising" is used when a metal such as high carbon steel which has been hardened and tempered, is heated to a bright red heat and allowed to cool very slowly. This turns something like a drill bit or a file back into a softened state, it can now be machined, drilled, sawn etc.

The same process can be used with our cast iron. It can be left in the sand until it is perfectly cool which will help, or remove it from the sand as soon as it has just solidified and place it in the furnace to cool down slowly. There must be enough space in the furnace for both the casting and the crucible; if the crucible is left out to cool it would probably crack from thermal shock.

Alternatively re-light the furnace when all has cooled down and place the casting inside and heat to a bright red heat all through. Then shut it down again to cool down slowly.

Something used in industry is Ferro Silicon, this is added to the molten metal and replaces the silicon lost from the iron during the melt, this is now available in small quantities suitable for the home caster. If you have any friend's crazy enough to be involved in this strange hobby, it might work out cheaper to buy a larger quantity from an industrial supplier and split the costs.

Plumbago, otherwise known as graphite powder is used in various ways in casting, often with cast iron, dusted into the mould to help the metal flow, fill some of the gaps in the sand, and help prevent the hard skin developing on the metal, coal dust apparently can be used in the same way.

Plumbago can be bought in an alcohol solution which is spread in the mould and then set on fire. It is worth considering drying a mould for cast iron, either by placing the whole thing in an oven, using a gas blowtorch inside the mould, or burning alcohol etc. This will help the metal flow and cut down on the steam and gasses produced to prevent gas bubbles forming in the casting, also to help with the hard skin.

The sand immediately around the casting melts and contributes to this hard skin, an aluminium casting will brush off nice and clean when it's removed from the sand, with iron sand will be imbedded on the surface. You can buy wire brushes to go in angle grinders which are very vicious; they look like they have wire rope in them, not the small needle like bristles in an ordinary wire brush. These will remove the worst of the rough skin finish and leave you with a semi-polished finish, this is called "fettling" and is very important before you try to machine a casting, both to keep the sand from the machine bed and to give the tool bit a chance, remember to take a deep, slow cut to get under the hard skin without overheating the tool bit. Another thing to try is to make the casting a bit bigger and grind the surface off before putting it into a machine for finishing.

One last thing while we're talking about cast iron, fluxing the metal. The slag from cast iron is a horrible, sticky, gungy mess, and takes some getting out. Lime has been used for years, put some crushed limestone in with the metal and although it creates more slag, it is much thinner and mixes with the iron slag to make it easier to remove. I've tried lime bought from a garden centre and this was very corrosive to the crucible so I haven't used it since. I rely on giving the metal a good stir, scraping around the inside of the crucible to dislodge any stubborn slag and then scooping it out. I tend to do this in stages, so when I lift the pot from the furnace it only needs a finishing skim before pouring, the less time between lift out and pour, the less the heat loss. There are commercial fluxes available which I might try one day.

On this subject, there are various fluxes, additives, de-gassing pellets available from suppliers for different metals, look at what's available and what you might need, don't rush out spending money until you know you defiantly need them, cheap and simple is my motto, but don't be too tight and spoil things!

How do we know when a metal is hot enough to pour? I've never looked at buying something suitable to measure the temperature of anything I've melted, lack of budget and I don't feel I need anything. Worst case scenario, I have to re-pour a casting that didn't fill properly, I will curse but it won't cost me enough to worry about. But it really doesn't take long to get the "feel" of the metal and pour successfully by experience. There is a simple and surprisingly accurate means of testing the temperature without having a clue how hot the metal is!

A must-have tool for home casting is a length of mild steel bar 1/4"-5/16" dia. about 3ft. long. By stirring the molten metal with this it will tell us a lot. Watch out for dross or slag, they will give you a false indication. Give the molten metal a quick stir with the cold bar, if there is a blobby mess on the business end then the metal is too cold to pour, if the bar comes out clean your ready to go, this seems to work with any metal.

I call this P.M.H.E. (plenty much hot enough!). However sometimes we need P.M.H.E.+, there are a wide range of pouring temperatures for each metal, a simple large chunk of a casting can be poured a lot cooler than something with a very thin section like a fan blade or something with cooling fins. That metal needs to keep hot enough to flow all around the mould while dissipating its heat to the damp sand as it passes through. With aluminium, wait for it to develop a definite pink/red sheen. With cast iron, the business end of the stirring bar should be starting to melt away.

Last but not least in this section are some more of the tools of the job. A bar of around 1/2"dia. 2ft. or so long with a taper on one end is good to tap the bod for direct melting. The bod stick I've made is crude but works, simply a large repair washer welded on the end of 3/8" steel bar with the edges bent over in a partial cup shape to help retain the clay, remember you also have a forge, put the welded washer in the furnace through the exhaust hole until nice and hot then it will bend easily with a pair of pliers (it's better to remove the metal from the furnace before trying to bend it!) use this method to make the tongs. Larger pieces can be held in the exhaust flame to the same effect.

In supermarkets they stock cheap stainless steel serving spoons, these make wonderful skimming devices, it's not recommended to rob the cutlery draw though! These cost less than a pound, stretch the foundry budget and buy a few. They will even work fine for cast iron although I only use them for the final skim once the crucible is out of the furnace. A large washer or small plate welded to a 2ft. steel bar serves for rough skimming in the furnace, I use a 2"x2"stainless steel plate about 3/16 thick with the corners rounded off welded to a length of 3/8" bar, this will need hitting with a hammer regularly to remove the slag.

My crucible tongs, like most of what I do were made from what was lying around. I have two lifting tongs and two pouring tongs for the different size crucibles I use, the picture show how they are put together.

On the lifting tongs, the parts that fit around the crucible are normally curved to the correct size, I have used flat bars. When the crucible has been used a few times, slag and muck builds up on the outside (maybe I'm just a messy worker!), so I thought these would create points where most of the pressure would be when lifting them. So gripping at four points should still give enough support to the crucible without damaging it, so far I've had no problems. This means that one pair of tongs will fit different size crucibles (within reason) and since I only use two sizes of clay/graphite crucible, everything else I lift is a metal pot it works for me. These do take up more space in the furnace when using them though so if your crucible has a large dia. you might be better off rounding these arms off.

A pouring shank is simply a ring of metal that the crucible can sit in at a suitable height to grip it well, welded to a suitable length shaft to hold it, these can allow the crucible to slip out when pouring. I have made pouring

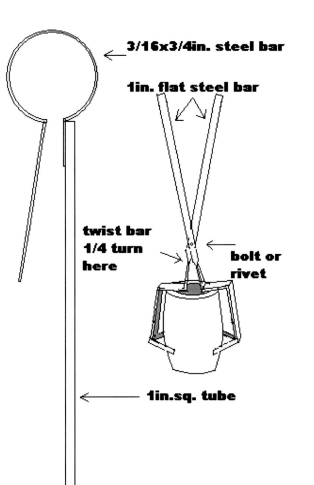

The headless man stirring the pot to loosen up the slag before skimming

The lifting tongs (in the drawing on the left) sit on top of the crucible to grip it in the right place

The pouring tongs are very slightly smaller than the crucible when squeezed closed

Lifting a crucible of cast iron

The healthy alternative veggieiron melted with used cooking oil

tongs which lightly grip the crucible to prevent this. The smaller one (my first design) works but I like the bigger one better. The ring opens enough to slip over the crucible from the top, although I normally place it down on the sand tray and place the crucible from the furnace directly in it. I squeeze the ring together and lift it until it is touching the crucible, then open my grip on the ring part a little, this allows it to go slightly higher up the crucible, now when I squeeze it together again the length of the arm on the ring part gives it enough "spring" to grip the crucible gently but firmly.

The long "pliers" shown are useful for gripping metal pots, placing pieces of metal in the crucible when it's in the furnace etc. I would be lost without them, I found mine dumped, but show a sketch how to make some. The flat bar the lifting tongs and the pliers can be made from are left over from a building project and can be bought at any building suppliers. The pivots for them are 6mm. bolts with two nuts locked together (to prevent them coming undone), and off cuts from the bar will be fine to weld on for the bits to fit the crucibles.

If you cast some iron crucibles with the lugs on the sides, or use the bolts sticking out from steel crucibles, a simple double hook arrangement can be used to lift them.

The long handled pliers

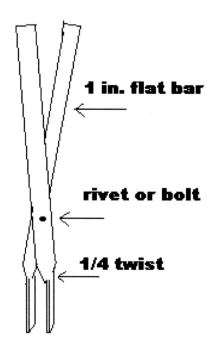

Casting

Of the many ways of producing a mould for casting, I'm only going to describe "green sand casting" in its most basic form; this book is about the furnace, not how to cast. But since several parts are cast I had better explain how to make them!

The "green" in green sand casting refers to the material being in an uncured state, not its colour. The green sand for our use is fine sand with a clay content to help it keep its shape when it's a little damp. This can be dug naturally from the ground, or we can blend our own.

Where I live there are no suitable shops to buy foundry supplies, so every thing comes mail order, this costs money, the heavier the order the more it costs. This would make a cheap item like a bag of sand expensive so I've always made my own. I've used play sand from toy shops, kiln dried sand for block paving and now I'm using silver sand mixed with a "secret supply". The finer the sand, the better the finish on the casting. I keep seeing silica sand mentioned on the internet but I can't seem to source any locally.

The story behind my "secret supply". We went on a long weekend camping trip to Lincolnshire. The weather was wet, the campsite depressing, it was quite a long drag to get there and we all became ill, probably from the incredibly warm, overcrowded swimming pool! There was an amazingly long beach (although it was too wet to spend much time there), with high sand dunes all along it. Only the lightest, finest sand gets blown to the top of a sand dune, I felt the quality of this stuff and all of a sudden this dismal trip didn't seem too bad after all! Now I'm sure there's a law concerning the removal of sand from a beach, however as you all know, you always get a little sand in your shoes when you go to the seaside, I did find it a bit awkward to walk back with 100kg. or so stuck in my shoes but it was worth it! I mixed this around 50/50 with my silver sand and am very pleased with the results.

The correct amount of clay in the sand is a matter of trial and error depending on which clay you're using. Casting iron fires the clay in the sand that comes in contact with it, which lowers the percentage of usable clay to bond the sand together making it necessary to "top up" the clay content regularly, I also add a little more wallpaper paste occasionally.

I have been meaning to try the clay in the soil where I live, it will need to be dried and then ground into a powder to enable it too be mixed with the sand, I'll get around to it one day. I have read of around 20% fire clay being used with the sand with reasonable results but I use Bentonite clay. This is a sticky clay which works very well, my mix is around 8% (by volume), look for pottery suppliers in your area, or John Winter's (listed in the back of the book) sell it. I also add some wall paper paste, around 1/2 of one of those sachets it comes in to 50kg. of sand. This isn't compulsorily, it will go mouldy and smell after a short time which some people find offensive (including my wife), but I do feel it adds a little extra bonding to the mix.

I have around 75kg. of moulding sand on the go all the time. I mix up 25kg. batches and then mix the lot together. It's best to spread it out on a plastic sheet if you can, the sand needs to be a little damp to mix it well, sprinkle the clay (and 1/4 of a sachet of wall paper paste if required) evenly over the sand and mix well. Once all of your sand is mixed it will keep indefinitely in air tight containers, or mine is kept in a sand box wrapped in a rubber sheet, I need to dampen it down every so often.

It needs to be quite dry, just enough moisture for it to hold together when squeezed in your hand without it sticking too much to your hand. This isn't a very helpful description, I'm afraid it's a case of using it and finding out for yourself when it's right. Do this testing with aluminium, the mould needs to hold together when the pattern is removed but not be wet enough for the metal to bubble when it's poured in, another sign it's too wet is the casting will have a rough, dull finish.

You need to get the moisture content right before playing with hotter melting metals! 8% of Bentonite clay will be about right, but you can play around a little until it suits you. Too much clay and the mould won't be porous enough to allow the gasses created by the hot metal to escape from the mould, although additional venting is always a good idea. Also if there's too much clay, the sand won't "give" as the metal cools and shrinks leading sometimes to fractured castings. The moisture content will need topping up with almost every mould poured as a lot will be lost as steam.

The clay should stay fine for a long time with aluminium, but hotter metals will fire the clay and more clay will need to be added from time to time. The finer the sand, the more venting it will need when it's rammed in a mould.

The idea of sand casting is to leave an impression or cavity of the desired shape in the sand which will allow molten metal to be poured in it. We can use a part we have and wish to reproduce or make a pattern from scratch to create the mould, but remember the metal shrinks as it cools so our casting will come out smaller than the pattern, generous rapping before the pattern is removed often is enough shrinkage allowance for a small pattern, otherwise the pattern needs to be made bigger to allow for shrinkage and any extra for machining to finish.

As a guide, allow around these amounts of shrinkage per foot length of metal, cast iron-1/10", brass or bronze-5/32", aluminium-1/4", although allow more for extruded aluminium. We aren't counting the costs in the same way as industry, so unless it's going to cause a problem with a particular casting be generous! The metals free, the fuels free and our times free (ok. I come cheap!).

Look at a bucket and spade for the seaside, the bucket is "tapered", we, sorry my kids fill the bucket with sand and pat it down firmly, turn the bucket over and rap the bottom to loosen the sand from the bucket and hey presto, a perfect reproduction of the inside of the bucket (well most of the time). We are doing exactly the same thing, except we need to leave a shaped hole not a pile of sand, the same sort of operation but reversed. The taper is a very important part of moulding, if the bucket had parallel sides, the sand would get caught and break up as the bucket was removed, we must use a taper on our patterns (known as draft) to enable us to remove them without damaging the sand cavity.

How much draft to give a pattern? I have a pattern book and it recommends 1/64" for every 1" of pattern depth, I tend to go as much as I can depending on what I'm casting. So the finished size of the pattern ready to ram up in a mould will be, the size you want the finished casting to be + the machining allowance + the shrinkage allowance + the draft allowance.

An open cast mould is where a depression is left in the sand like a footprint on a beach, and metal is poured in this depression. This has limited usage as the side open to the air will have a poor finish. It will probably shrink and leave a hollow in the middle, and with aluminium it will show a lot of shrinkage on the good side, and possibly not fully run into all areas of the mould as there isn't enough weight to press it into small cavities.

Mainly we will be using a closed mould where there is a top and bottom which fit together and the metal is poured through a hole called a sprue. The taller this sprue is, the more weight of metal is pressing down into the mould cavity, this helps push the metal into small areas. And if the sprue has a larger cross section than the casting it will take longer to solidify, therefore as the casting cools and shrinks the sprue acts as a reservoir to top up the metal in the mould. Sometimes two or more sprue's are used, the additional ones are called risers. These can help the metal flow by giving a good gas escape route, as well as adding a reservoir to areas that might shrink, or simply to allow two people to pour a mould at the same time, such as when one furnace can't melt enough metal for a large mould and two are used.

To make this work the sand needs to be put into a box so we can handle the mould without it breaking up, these are called flasks. A flask consists of two or more sections which can locate back together in the same place once they've been taken apart. For now we will deal with a two part flask, the bottom part is called the "drag", and the top is called the "cope". My mind's a bit soggy these days and I can't remember anything! So I think of it as, life is a "drag" when your at the bottom, but can you "cope" at the top.

Both the cope and the drag consist of a box with four sides and no fixed top or bottom, they need to have the same length and breadth but can be different heights, and a means of the two locating together. The easiest material to make them from is wood and I have drawn two ways they can be built. Screw and glue, or nail and glue them together, as they get larger it's worth fitting a rib around the inside to help hold the sand in. Allow at least an inch all round, better two between the pattern and the flask sides to stop the molten metal burning the wood.

Most of mine are made from rough sawn, unpainted wood; it seems silly to make something smooth and shiny only to have to do extra work to make the sand stay in place.

The type using the angle iron will allow for the top and bottom to be slightly different sizes (if your carpentry is as rough as mine!), and still locate correctly by drilling the second holes for the locating pins once the angle is screwed to the flask, and the two halves of the flask are clamped together. Leave enough gap between the two pieces of angle so they also serve as handles. But make sure you drill the holes nicely in line so the pins are parallel to allow the two halves to lift apart!

We also need moulding boards to go under the flask to stop the sand falling out of the bottom, if these are an inch or so bigger than the flask size (all around), it makes them easier to handle, something like plywood is fine. Two per flask are needed.

The fuel valve

The fuel valve is simply a block of aluminium with a 15mm.(1/2") copper pipe stub coming out of the top which connects to the ball lever valve, and a 10mm.(3/8") copper pipe stub coming out of the bottom to connect to the oil pipe in the air pipe. There is a hole connecting the two pipes internally which starts from the outside and has a thread tapped in it to allow a 10mm.(3/8") stud with a tapered end to act as a large needle valve. This tapered stud variably blocks the hole between the two pipes to adjust

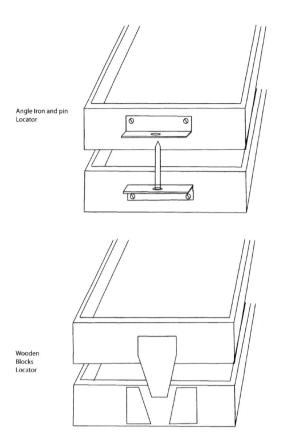

Angle Iron and pin Locator

Wooden Blocks Locator

A flower pot being used as a pattern for a crucible

The sketch to the left shows two ways to locate the cope and the drag, the top using angle iron and pins, note the pin should be in the cope not in the drag as shown.
The bottom using wooden blocks

The pattern removed, ready to close up and pour

Pouring cast iron into the mould

the fuel flow. There is also a bleed valve which allows the pipe work and valve to drain when the fuel is turned off from the ball leaver valve.

To cast the fuel valve, make the wooden pattern as shown and seal it by painting/varnishing it (some of these fast drying aerosol paints are handy if you are in a hurry). Cut the copper pipes to about four inches long and tin well with solder where they will go into the aluminium to help them seal (yes you can get lead solder to stick to aluminium). Ram the pipes full with moulding sand to prevent the aluminium filling them up and then put them in an oven on a gentle heat to dry the sand to help prevent gas bubbles forming in the mould.

Place the drag upside down on a board, I use a gardening soil sieve to sieve the sand in the flask this ensures the sand is evenly distributed over the area with none compressed together. Sift a couple of inches of sand over the board and then begin to ram the sand in place using something like an 1"-1,1/2" dia, length of wooden dowel. Evenly ram the sand down into the edges of the flask and across the board, use a diagonal action to make sure the sand is compressed into the sides of the flask.

Once the bottom is covered with sand, the rest of the drag can be filled with sand without sifting it but still ramming it in position, and then a straight edge can be scrapped along using the wood of the drag as a guide to finish off flat.

Unfortunately the amount the sand needs to be rammed is a matter of experience, too hard and gasses can get trapped in the cavity and when the casting shrinks it can fracture if the sand has no "give". Not hard enough and cavities can be left in the mould surface giving a defective finish, this can look as if your sand is too coarse because the grains haven't been "shuffled" around enough to fill all the possible gaps, or the sand will not sit in the flask properly and can move slightly damaging the mould cavity.

Now put the other board over the drag and clamping the drag between the two boards with your hands, turn it all over. Remove the top board to reveal a nice flat sand surface. Put the tinned ends of the pipes into the pattern, place the pattern in the middle of the drag and gently and evenly knock the pattern and pipes into the sand until they're half way down. If the sand around the pattern and pipes now dips down, carefully add some sand to these areas and smooth it in with a suitable small metal tool.

Assuming all's gone well first time (I'm an optimist!) we need to sprinkle parting powder evenly over the sand and pattern in the drag. Shop bought parting powder is worth buying (John Winter's sells it), however an easily found alternative is talc, it works but you need to use more than you would with parting powder, (on your own head be it if you take it from her highly

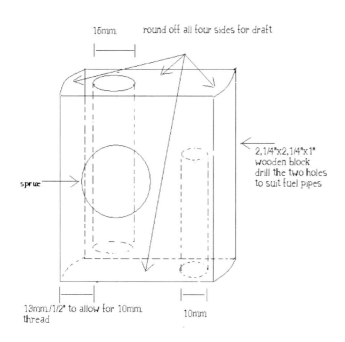

The wooden pattern for the fuel valve

The pattern rammed up in the drag

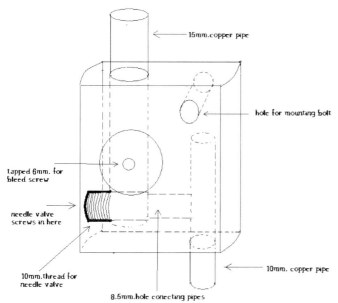

Detail to finish fuel the valve

The finished fuel valve components

expensive, designer wiff range in the bathroom!) Watch out for getting a build up around the pattern. The parting powder is to stop the next layer of sand we ram up in the cope from joining forces with the sand in the drag forming a solid mass; we need to be able to separate the cope from the drag to remove the pattern.

Now I'm really going to risk the displeasure of your other half! They have very soft gentle brushes they use to apply powder to their faces; these are exceptionally good for moving powder around our moulds without damaging the sand, getting excess from around patterns etc. a few of these in different sizes are invaluable.

Back to our mould, put the cope in place, we now need to think of how the metal is going to get in the mould, the sprue will be formed with a piece of wooden doweling, a piece of plastic or metal pipe, anything that will leave a nice hole to pour the metal in, about an inch in dia. is good. This needs holding in place on the pattern as shown above the 15mm. pipe (when the sprue feeds directly into the cavity like this, it is referred to as a pop gate). Now sprinkle sand in the cope and ram it in making sure it is firmly rammed around the pattern and the dowel that will form the sprue until the cope is full. I use my hand to press the top down to firm up the sand on the surface.

Rotate the sprue dowel and pull it carefully from the sand, using a finger press the sand around the top of the hole to loose the ragged edge and form a slight funnel. Any loose sand here when the metal is poured will wash into the cavity and mar the casting.

By lifting smoothly and straight up, remove the cope from the drag, and rest the cope on its side. If a little of the sand around the pattern has broken away, it doesn't really matter as a quick file afterwards will soon remove the excess metal. With a finger again, gently firm up the sand where the sprue enters the cavity. Using a straight piece of 1/8" wire, make a few vent holes by carefully pushing the wire from the inside to the outside of the mould.

I use a very sharp brad awl; tap the point into the centre of the pattern, and then gently tap the awl from all around until you can see the pattern and pipes moving in the cavity a little in every direction, this is called "rapping" the pattern. Remove the awl and carefully screw a small hook or eyelet in the hole it leaves, it only needs to go in deep enough to allow you to lift the pattern out, which you do carefully!

If there is an area where the sand has come loose but not fallen away, a small, wet paint brush touched on the area often will allow it to bind together again. I have a rubber bulb which is for blowing small debris away I bought

The finished fuel valve

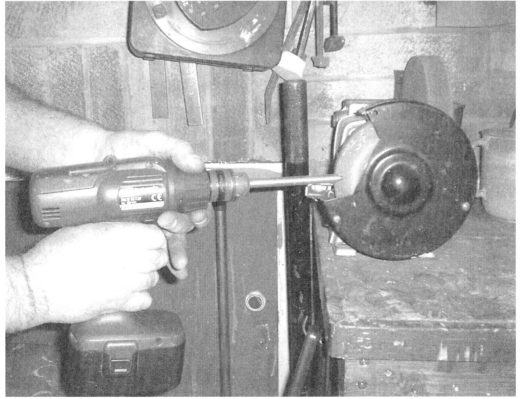
Grinding the taper on the "needle valve

from a model engineering shop. This is very useful for blowing any stray particles of sand out of the mould, if not take a deep breath and blow!

Carefully put the copper pipes in place in the sand and replace the cope, this is now ready to pour. Put the mould near the furnace and melt some aluminium and fill the mould with metal. I always have either some DRY bricks on hand or some of those kitchen tins used for small cakes or small Yorkshire puddings to pour the excess metal into for ingots.

You should always allow the metal to cool right down before removing it from the sand. Different parts of the casting will cool at different rates, this will cause the casting to distort and build up stress. By leaving it in the sand, it's insulated and cools much more evenly. However, with something like this it doesn't really matter if you rush things a bit! I'm always impatient to see how things have turned out.

Poke and wash out the sand from the pipes and cut the sprue off leaving a stub around 1/4" high, file this as flat as you can and clean up any "flash" (this is metal that has crept out around the parting line).

Drill the connecting hole between the two pipes, mine is 8.5mm. the tapping size for the 10mm. thread, and tap the thread in the area between the outside and the 15mm. pipe only. Then drill and tap the 6mm. hole through the sprue stub into the 15mm. pipe for the bleed valve. I cut the copper pipes shorter, leaving enough left for the compression fittings to go on.

The "needle valve" is just a piece of 10mm. studding with one end ground to a blunt taper. I put the studding in a drill and held the spinning stud against the wheel of a bench grinder until I'd ground a taper reducing the end by about 2/3rds. and back up the stud for about 10mm.

A knob or some means of gripping the stud in use is needed, two nuts run up the stud and locked together will work, or trap a large washer between the nuts to give a better grip, or if you have a suitable knob use it. If you run the stud into its hole and bottom it out a few times it will form a "seat" for itself. Blowing down one of the pipes will let you know how well it's seating, it doesn't need to be perfect but if you ever use thin oil it's nice to be able to slow the flow right down if necessary. To seal the thread to stop oil creeping out, either wrap some P.T.F.E. plumbers tape around the stud, or a generous smear of non-hardening gasket cement will do the job.

The bleed screw is a 6mm. (1/4") screw or bolt short enough to go in its hole without bottoming out. File 1/3rd of the thread off to leave a flat the whole length of the threaded section, this is so that when the screw is loosened air can bleed down this flattened area into the valve, a rubber O ring sitting between the filed off sprue and the screw head seals this off when the screw is tightened up. Once again a knob is needed, slip a large washer on the screw, run a nut up to trap the washer, then fit your O ring. Don't over tighten this as the O ring will squeeze out.

Fit the fuel valve with the compression ball lever valve on the 15mm.pipe, a short stub of pipe the other side of the valve will connect it to the flexible hose, and a compression straight coupling on the 10mm.pipe to connect it to the pipe coming from the air pipe.

The fan housing

Sometimes a casting is needed and the pattern can be quite involved to make, if several of these castings are needed then it might be worthwhile to do the job properly. However if you only need one and it doesn't matter if it's a bit rough, then lost foam casting can be the way to go. Basically foam is cut to shape and carefully rammed in a flask, I've heard of people pouring just dry sand around a pattern although I would always use moulding sand. The sprue is a pop gate, i.e. leading directly to the pattern, not via a gate, and the mould isn't opened to remove the pattern. When the hot metal is poured into the sprue, the foam vaporises and the cavity is filled with metal.

There are all kinds of synthetic foams available, choose something that is cheap and available and won't poison you with the fumes! This should always be done in the open air, I always break the casting from the sand in the open air as well as the smell penetrates the sand, I leave it outside to disperse a little.

This is a fun and fast way to reproduce a shape and since the pattern is not removed from the sand, very complex shapes can be produced, especially if several pieces are glued together.

For the fan housing I reached deep into my pocket and purchased the latest in High Tec. casting materials and bonding substances. A box of polystyrene ceiling tiles, a reel of sellotape and some long thin nails (dress making pins should work well). Actually I have a large box of ceiling tiles in the loft that were given to me and I had the rest in the garage anyway!

For all the fans I've made I've always used blades that are commercially produced and off other things, i.e. the one shown for this fan is from a flymo lawn mower, others have come from clothes driers. I make an aluminium copy of the fan to use so I have the original as a pattern for future reference. This allowed me in the case of the flymo fan to add a centre hub by using an 1,1/2" sprue in the centre of the plastic pattern, and cutting it to length afterwards so I could drill the centre hole of the casting to suit the shaft I wanted to mount it on, and use a grub screw to hold it in place.

If you start with this furnace using a hoover blower you will know how much air it will need, if you already have a blower from a previous furnace then try it. Centrifugal fans from other appliances are readily available for nothing if you scavenge around, if you decide to cast a copy of one, make sure the metal is very hot to fill the mould and rap the pattern very well to get it out.

A 2800 or 3000 rpm. motor is needed, and on my one I've belt driven between the fan shaft and motor giving a 2-1 increase of fan speed, although a larger fan would probably work fine straight off a 3000 rpm. motor, this flymo fan is very thin. Remember my motor is only $1/10^{th}$ of a horse power and comes from a spin drier, although the no load speed is 2800 rpm. the fan is probably only turning at around 4000 rpm. due to the load on the motor. I have used a stretchy 1/4" belt to lessen the friction and not have to bother with any form of belt adjuster, (this stuff is available mail order from Chronos listed in the back of the book or try model engineering suppliers). The belt is cut to length and the ends are heated to weld it together (very simple no specialist tools needed, ask for instructions when you buy it).

How you want to mount the motor and drive the fan I shall leave to you, it largely depends on your fan size and motor size, use the pictures of mine as a guide if they help.

Making the fan housing is very quick and easy. There are professional ways to "develop" the expanding shape of a centrifugal fan, you must have realised by now, I don't "do" professional I bodge! My polystyrene tiles measure 300mm.(11,3/4") square, and my fan measures 210mm.(8,1/4"). Lay the fan on the tile as shown and carefully draw freehand the expanding, snail shell shape around the fan with a felt tip pen, remember to mark where the centre of the fan is. Cut this shape out using a very sharp craft knife or a hot knife, if your fan is too big for this to work with one tile sellotape another one to it.

Depending on how big a hole you will need for the shaft for the fan, on mine the centre hub sticks out through the housing, cut out the centre hole as required. Place the tile with the fan on it on a flat surface and gauge the height the sides will need to be. Measure from the flat surface to the highest point of the fan. With the best will in the world, this housing won't come out of the sand perfectly flat, so allow plenty of additional clearance add 1/4-3/8" or so to the measured height. Cut some strips of tiles this wide, and starting from the tightest part of the "snail shell" start to fit the side, the first part will jam in the corner, use small pieces of sellotape to hold the side in

This shows the blower in action, note the clay lined pre-heating in the exhaust flame

The curved blades of the fan and cast front cover

The 2/1 increase belt drive

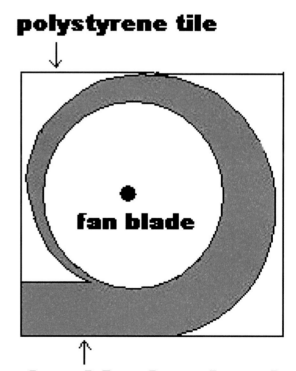

The idea for drawing and cutting out the polystyrene pattern

position as you go around. When you need to join the next piece on, push a couple of pins through from one piece into the other to strengthen the joint, and sellotape well, it won't be perfect but a bit more adjustment is possible when it is rammed in the sand, the pins stay in the casting.

Sometimes a few pins stuck through the sides at an angle into the main part will help to hold it in shape, fill any gaps by laying sellotape across them. Cut and fit the last small piece of side which goes from the corner to the outlet and cut a small piece of wood to the size of this outlet rectangle and temporarily tape it in place to hold the outlet in shape, I forget this every time and all my outlets are misshapen! Once most of the sand is rammed into place and will act as a guide to hold the polystyrene, remove this wood and fill in the sand.

Place the finished pattern on a moulding board so the sides are sticking up and put the upside down drag around it. Start filling up the sand, I only use my hands to "ram" this type of mould. Carefully pack the sand both inside the sides and out to maintain or alter for the correct shape until the flask is full and levelled off. Put another board over the sand and turn the mould over, don't use any parting powder, put the cope in position. Judge the centre of the casting and place a good sized dowel (1,1/2"dia.) there for the sprue, I always use two risers (about 1"dia.) with these as well, one on the far side of the outlet section and the other on diagonally opposite side of the pattern. By filling the cope with sand from the area of one riser they will hold themselves up as you go along, the reason for the risers is to allow the gasses to escape easily, I also usually poke a few additional holes in the sand with wire, however it is difficult to know when you reach the polystyrene, my last one came out with needles sticking downwards from the casting and dimples on the top. When you remove the sprue and riser formers try to blow as much of the loose sand out of the bottoms of the holes, a damp paint brush can lift sand particles up as well.

Get the aluminium good and hot and pour as fast as it will go in the sprue, it should come up the risers if all goes well! When the casting comes out of the sand the areas where the sellotape was will be discoloured, this is only surface staining and will wire brush off.

As the housing will be slightly out of shape, place it on a flat surface and file off the worst of the high spots where the front cover will go, I don't go to mad with this as a little silicone fills a lot of gaps!

To make the front cover, place the housing on another polystyrene tile and draw around the outside, drop something down the fan centre hole to mark this as well. Cut out the outside shape (remember the metal will shrink, so cut to the outside of the marked line to allow for the shrinkage), and use a compass to mark out, and then cut the centre hole to suit your fan. Ram this up the same as the housing and pour.

If you are unsure of the correct sized hole, make up the rest of the blower and make a temporary cover (or covers) from cardboard or hardboard and hold them in place, altering the hole size until the best size is determined, then cast the proper cover.

To fit the front cover on I used a bead of silicone, and drilled and fitted a few self tapping screws through the cover into the sides.

To connect the blower to the air pipe I used a piece of thin steel sheet and bent it to suit the rectangular outlet of the blower one end and the circular air pipe the other using a gentle reducing taper. I welded the air pipe to this connecter, and used self tapping screws to attach it to the outside of the fan housing.

A household light dimmer switch controls the speed of my fan; the picture of the other fan shows an adjustable shutter over the air inlet to control the air flow, this fan is directly mounted on the motor shaft.

Cast iron crucibles

These are easy and cheap to make and are better than welded pipe crucibles. They are quite heavy, but the extra weight means more mass to retain the heat.

The largest one I have at present holds four pints; I think this is probably as large as I will go due to the weight, unless I find a suitable pattern which is thin enough to lighten things up. Since I have larger steel crucibles, such as cut down fire extinguishers and a re-welded lawnmower roller, I'm not too worried about it!

The patterns are varied, mostly I've used clay flower pots. Try to find thin walled ones, they need painting as the clay is porous. However I found a nice "crucible" shaped tea mug which holds 3/4 pint, just right for brass and bronze. A quick attack with an angle grinder to remove the handle and an instant pattern! Have a look around, what looks good? As long as it has draft and isn't extremely thin so the metal will have trouble flowing into the mould created by it, give it a go!

I have used the tea mug as it is, but will be drilling a hole in the bottom of it to help get the sand out of the middle (sand or anything else in a mould which is there to leave a hole in the casting is called a core). A thin metal plate placed in the mug before the sand goes in, will allow a rod to be pushed through the drilled hole and against this plate to help get the core out in one piece. Flower pots either have a hole or one can be drilled.

The flask only needs a shallow drag but a tall cope for the height of the pot. Rather than build special flasks for these, I usually build the sand up over the height of the cope frame. An alternative is to build an extension piece which will fit onto your cope to extend its height.

Place the metal plate in the bottom of the pot and ram it full of sand, now ram up the drag and using the moulding boards turn it over. Place the sand filled pot upside down on the drag and push it down firmly $1/8^{th}$" or so, dust with parting powder and fit the cope.

Ram up the cope ensuring tight ramming against the pot, until the pot is an inch or so under the sand. Using a finger, carefully clear the sand from above the pot, only as wide as the bottom. Press this sand to a firm finish so none falls in the mould when poured.

Using a small piece of wood as a drift against the exposed bottom of the pot, rap gently until you see the pot moving slightly in all directions. Lift the cope off and lay it on its side, tap the pot on its side all around with a piece of wood and using a rod to hold the metal plate, carefully lift the pot away.

You will need to tidy the sand core a little when you remove the plate, then blow away any excess sand from all the mould and refit the cope. Pour the cast iron until it covers the core about the same amount as the clay pot.

If you want lugs at the top to lift crucible and a tongue to grip for pouring at the bottom, then before you dust with parting powder, place two pieces of doweling against the pot diagonally opposite each other, and push them half way in the sand, now ram up the cope. When you've cleared the sand from the bottom of the pot, before you rap it, cut away an area that the metal can run into to form the tongue, 90 deg. from the lugs.

You can mess around fitting pouring spouts to your flower pots with filler if you want, I'm too lazy!

One thing, I get about a 20% failure rate with these, it's easy to tell if the metal wasn't hot enough, but sometimes I get a hole for no obvious reason (not slag inclusion). Certain sizes/shapes are worse than others, I'm putting this down to gas rising up from the core and forming a gas bubble. I will try drying the core, but since I have plenty of crucibles and am busy sorting this book out, it will have to wait, just thought I'd let you know.

The following is a list of companies I've dealt with, books I've found useful and a few web sites I like.

John Winter and Co. P.O. Box 21 Washer Lane Works, Halifax, tel. 01422 364213,
 Most foundry supplies, especially useful as they have a model engineering dept. for smaller quantities.

Hoben int.ltd. Spencroft Rd. Newcastle-under-Lyme, Staffordshire, ST5 9JE
Tel. 01782 383000,
www.hoben.co.uk Lots of High Tec. casting supplies, good prices for Salamander crucibles.

Purimachos, Waterloo Rd. Bristol, BS2 OPG. Tel.0117 955 4361 Refractory manufactures, I have found them very helpful and will always buy my refractory from them.

Chronos Ltd. Unit 14 Dukeminster Estate, Church St. Dunstable, Bedfordshire LU5 4HU,
 tel. 01582 471900,
www.chronos.ltd.uk Engineering Supplies, worth getting a catalogue.

Books

Dave and Vince Gingery www.gingerybooks.com, I will never hesitate to buy a Gingery book, I might not follow their designs exactly, but there is too much information in them to ignore if the title is for your interests.

Foundrywork For The Amateur and The Backyard Foundry by B.Terry Aspin . Two books that overlap each other, but I'm still glad I bought both of them. Great information on pattern and core making, mainly for model engineering, but casting is casting!
www.specialinterestmodelbooks.co.uk

Camden Miniature Steam Services, Barrow Farm, Rode, Frome, Somerset BA11 6PS.
 tel. 01373 830151
www.camdenmin.co.uk
  Not a book but a book company I have bought a lot of books from, a wide range of subjects covered, and good service

Check out

www.backyardmetalcasting.com

www.foundry-fopars.co.uk